Kong, G
and the Living Earth

Kong, Godzilla and the Living Earth

*Gaian Environmentalism
in Daikaiju Cinema*

ALLEN A. DEBUS

Foreword by J. D. Lees

McFarland & Company, Inc., Publishers

Jefferson, North Carolina

ISBN (print) 978-1-4766-8721-6
ISBN (ebook) 978-1-4766-4653-4

LIBRARY OF CONGRESS AND BRITISH LIBRARY
CATALOGUING DATA ARE AVAILABLE

Library of Congress Control Number 2022024605

Front cover illustration © 2022 CSA Images

Printed in the United States of America

*McFarland & Company, Inc., Publishers
Box 611, Jefferson, North Carolina 28640
www.mcfarlandpub.com*

To the memory of Edna P. Debus,
George W.W. Debus and Justina Lopez Garcia

Table of Contents

Foreword
by J. D. Lees

My nearly lifelong fascination with King Kong and Godzilla, while admittedly a bit out of the mainstream, is probably not atypical for a Monster Kid of my vintage. I was born in 1955, sandwiched between the Japanese release of the original *Godzilla* in 1954 and its Stateside debut (as *Godzilla King of the Monsters!*) in 1956. However, unlike most young children of more recent decades, I have no memory of exposure to a dinosaur of any sort for the first six years of my life. Not that they weren't around; at the very least they entered our home with the weekly "colored funnies" in a newspaper comic strip called Alley Oop, one of several that my father enjoyed.

It was in my Grade 1 classroom library where something happened that has stayed with me to this day. I discovered a book called *The Wonderful Egg*, written and illustrated by the imaginitive American artist Dahlov Ipcar. The book told of a mysterious antediluvian egg and then speculated as to which of twelve prehistoric reptiles it might belong. For reasons I still do not fully understand, I was instantly captivated by those amazing creatures that existed before the time of man, a fascination that has never left me.

Today, an interest in dinosaurs, by child or grownup, is rather commonplace; after all, dinosaurs are ubiquitous in current culture. By contrast, in the early 1960s dinosaurs were widely known but considered to be largely irrelevant remnants of a bygone era. First identified in 1842 by Sir Richard Owen, the fabulous beasts generated considerable excitement during the late nineteenth and early twentieth centuries, as their fearsome fossils were unearthed and mounted in museums around the world. By midcentury, however, they had become somewhat old hat, evolutionary losers that grew too big and too stupid to survive. Thus it was that I found the adults in my childhood world amazed that I, at

1

the age of six or seven, could pronounce the tongue-twisting names of long-extinct mammoth reptiles. (Now, in the age of *Jurassic Park* and its many sequels and imitators, any two- or three-year-old can perform the same feat.)

It would be interesting to pinpoint what it is that makes only certain people become fascinated by dinosaurs. Several of my contemporaries had dinosaur books or plastic toys, but nobody cottoned to them like I did, to the point that my Grade 5 teacher asked me to take over and teach the class when we undertook a unit on the dinosaurs and related reptiles. Perhaps there is a recessive "dinosaur gene" tucked away in the human genome, or maybe the matter is best left unresolved. Whatever the cause, there's a distinct subset of humanity that can't get enough of the long-gone behemoths.

Not long after my initial dino-exposure I discovered an amazing new dinosaur, far bigger than all the others, and rampaging through the cities of the modern world. *Godzilla King of the Monsters!* was broadcast in all its black-and-white glory one day on our local 4:30 p.m. Blockbuster Movie. Again, from the moment I saw the titular beast, I was hooked. Though I wouldn't go so far as to describe myself with the Japanese word "otaku" (meaning "obsessed fan"), an impartial observer might disagree. Growing up, I would never miss the opportunity to watch (or rewatch) a "G-movie," whether first-run in the theater, on TV, or at that lost and lamented Saturday afternoon ritual, the Kiddy Matinee.

At the time there was no such thing as an action figure, but my Aurora models of Godzilla and Kong proved ample substitutes, fighting many a sandbox battle. I even duplicated Godzilla's fiery ray by placing cut-off firecrackers to shoot flames from the model's mouth, as the "battle-damaged" figure's burn marks bear witness to this day.

Reaching adulthood, I found my place in the world as a husband, father, and secondary school teacher of science and math. Concurrently, and thanks to the invention of desktop publishing, my pleasant and occasional "monstrous" pastime evolved into somewhat of a second job as the editor and publisher of a quarterly Godzilla journal called *G-FAN*, and the convener of an annual convention of giant monster enthusiasts called G-FEST. I must admit it: Godzilla has been thoroughly interwoven into my life.

King Kong holds a different, though no less special, place in my heart. My dad, being born in 1920, had seen the 1933 film during its original theatrical release, and it was he who introduced me to the Monarch of Skull Island. I don't remember the exact date, but it was shortly after my discovery of dinosaurs and Godzilla. I do vividly remember the occasion and the setting. We watched a late-night broadcast of the movie on the TV in our family room, my mother and sisters having retired for the night. Having since become a father myself, I can well imagine my dad's feelings, watching his son drawn into a fantastic world populated by the dinosaurs he'd heretofore seen only as static images in books, now lumbering and bellowing and engaging in mortal combat. I remember still the amazing reveal of King Kong, the non-stop action to the movie's climax, and my genuine sorrow as the gallant anthropoid was cut down by the buzzing biplanes that tormented him atop New York's Depression Era skyline.

Throughout the course of my life, both Godzilla and King Kong have gone on to rack up impressive filmland careers. In this category, Godzilla is the undisputed champ, with 33 live-action movies and several animated adventures. Kong's appearances have been more sporadic, but include seven feature films and an animated TV series. However, there have been a number of movies, like *Konga, A.P.E.*, and *The Mighty Peking Man*, that have featured a giant anthropoid intended to capitalize on the fame of the genuine article. Only twice have the two preeminent titans encountered one another onscreen: in the Japanese classic *King Kong vs. Godzilla* (1962), and nearly six decades later in *Godzilla vs. Kong* (2021).

Add in all the ancillary paraphernalia, the books and toys, comics and collectibles, the countless media references … and it's crystal clear that Godzilla and King Kong are every bit as powerful as cultural forces as they are the unstoppable engines of destruction that they appear as onscreen. Not everyone shares the depths of my devotion to them, but they have undeniably captured the recognition and interest of a global audience far beyond the reach of any other giant monsters. The question is: why?

On a superficial level, Kong's appeal is easier to understand. He is humanoid, and therefore relatable to audiences. Filmmakers have consciously striven to make him sympathetic, imparting unto him an almost human personality, while at the same time showcasing his traditionally manly virtues such as power, confidence, bravery, and protectiveness of the women that have caught his fancy. Godzilla's case is more complex. Though more wantonly destructive than King Kong, Godzilla is actually a casualty of human carelessness, revived and radioactivated by the cataclysmic nuclear tests in the South Pacific. Not as humanoid in appearance as Kong, Godzilla nonetheless has been allowed to show flashes of both intellect and personality, while his many battles with other kaiju have showcased such commendable qualities as bravery, determination, and grit.

The long fascination of movie audiences with monsters and creatures is undeniable. In that very select group known as the "Classic Monsters," Godzilla and King Kong stand apart. Ironically, the human monsters like Dracula, the Mummy, and the Wolf Man offer little beyond unrestrained evil, whereas the animal monsters Godzilla and Kong incorporate admirable human qualities to which an audience can aspire. This facet of their characters is part of what has made them memorable, likeable, and, ultimately, successful.

Is that all there is to it? Can Godzilla and Kong's position atop the pantheon of cinematic creatures be the result of mere likeability? Methinks not! To achieve such heights of notoriety and popularity, the fascinating phenomena that are King Kong and Godzilla probably have much deeper connections to the human psyche and the human condition. Dare we consider an even more primal and subliminal link, one that spans vast stretches of time and is planetary in scope? If the thought intrigues, you have the right book in hand, for no one is better qualified than author Allen Debus to explore, explain, and usher readers through the intricacies of King Kong and Godzilla's relationships not only with each other, but with humanity and the very Earth itself.

I've previously extolled the strengths of Allen as a devoted student, researcher, and chronicler of both cultural and natural history, focusing especially on dinosaurs, dino-monsters, and kaiju, and their place in and influence on popular culture. He has written over forty articles for *G-FAN*, and I've acted as his editor on every single one, so I know his work. I've been impressed not only by the breadth of his knowledge and his prodigious memory, but maybe most of all by his amazing ability to synthesize disparate information and ideas into coherent commentary. In recent years he's devoted considerable time and energy to considering the Godzilla/Kong/Earth connection, and weaving its various (and sometimes unconventionally obscure) threads into a bright tapestry that he's offering, through this book, for your consideration. Agree with his conclusions or not, I guarantee you will be challenged, intrigued, inspired, and left with a new and even greater appreciation for those two now-and-forever Kings of the Monsters, Godzilla and King Kong.

J. D. Lees, a retired educator who lives in Manitoba, is a devoted fan, popularizer and researcher of Godzilla, and Kong. He organizes the annual G-Fest held in Chicago, and biennial G-Tours of Japan, and edits the magazine G-FAN.

Preface

A Future of Biblical Proportions

As actress Gillian Anderson, star of *The X-Files*, once said, science fiction is apocalyptical.[1] Don't we increasingly find ourselves living in a *Twilight Zone*–ish, science fictional state of existence, an intractable mind cage that seems ever more apocalyptic? One where toilet paper is so treasured we'd rather not flush it down?

We do live in an apocalyptical time, one metaphorically captured by Legendary Pictures' string of giant dino-monster, daikaiju films of the 2010s and beyond. The possible, plausible extinction of mankind and demise of civilization have become prevalent themes in popular culture—no less so in Legendary's genre films. And yet theories of a living super-organism, protective planet "Gaian" Earth are reflected in these films, like never before. Protective of whom or what, though? Or is Earth instead "Medean," poised to eradicate species superseding natural bounds? Is our planet out of balance?

This book considers the concept of nurturing Mother Earth's Gaian principles as projected in the Legendary series and tie-in novelizations. King Kong and Godzilla play central roles in Legendary's new, ecologically-themed millennial series; in fact, their immortal conflict is primordial ... ominously apocalyptical. On a highbrow level, esteemed microbiologist René Dubos stated in 1959: "Human destiny is bound to remain a gamble, because at some unpredictable time and in some unforeseeable manner nature will strike back."[2]

Formerly we referred to such rare yet seemingly insurmountable conditions as hellish or of biblical proportions. But our science and means of understanding nature have moved on since then.[3] How sobering to realize that we're living on a planet about to go way apocalyptic, where its alleged "ruling species" could be self-annihilated. Too often scientific warnings have been ignored, which is why circumstances today are approaching the apocalyptical instead of just seeming bib-lical—an old-timey term that in light of present Gaian circumstances simply sounds archaic. An outcome of an increasingly secular soci-ety? Or instead, culturally, maybe we aren't secular enough to right-fully resolve such dire problems? We've been lucky before—for example, avoiding nuclear war with the Soviet Union when it seemed imminent during the Cold War—but how lucky do we feel now with so much at stake environmentally, geopolitically and with a pandemic at large? Certainly, it's time to shed the scales from our eyes.

"Gaian"? Yes—this is the new understanding of our living planet arising from 1960s philosophical and scientific meditations persisting today in the form of Earth Systems Science—reflecting the totality of biological and geochemical interactions that have made Earth a habit-able world for billions of years. The current geological epoch in which we live has been pronounced the Anthropocene, reflecting Man's indus-trial impact over the globe and an ongoing human-driven Sixth Extinc-tion event (while further acknowledging five mass extinctions of the fossil record, including one that extinguished the non-avian dinosaurs 66 million years ago and another, the Permian "Great Dying," 251 mil-lion years ago, to which 98 percent of all life succumbed).

So, besides scarcity of toilet paper, what are these dire problems mankind faces in the 21st century ... as if we don't already know. They're in the news every day: it all seems overwhelming. In January 2020, prophetically given "existential threats of climate change and nuclear weapons," the Doomsday Clock's minute hand was "moved closer to midnight *than ever before*."[4] The prospect and grim reality of climate change is debated politically in efforts to dismiss, although without tackling, our escalating crisis.

During the cresting of the coronavirus pandemic, many were living like there was no tomorrow. (Do "they" know something that generally hasn't dawned upon the rest of us yet?) Is America's general reaction and response to COVID-19 a microcosm for why and how our current civ-ilization might not survive the present eco-environmental crisis? And despite all efforts there are still thousands of nuclear missiles on the planet, ready for launch. In the midst of all this, let's not exclude the ongoing social crises inflicted upon peoples of the world.

Then, while I was writing chapters for this book, along came COVID-19.[5] I enjoy knowing that I'm only two handshakes removed from "touching" my paleoart[6] dinosaur artist hero, Charles R. Knight (who passed in 1953, 16 months before my birth). Although present cir-cumstances may force death of the traditional, congenial handshake.

For the record, while I do find the nature of catastrophes of the geological past intellectually intriguing, and find fictional, metaphorical tales of global or cosmic cataclysms thrilling.... I do *not* revel in the chaos of present reality. It's like our love-hate relationship with Hollywood monsters and horror movies; we love to feel scared at times while reassured that we're really safe: after all, it's *"only* a movie." Yet in cases where movie messaging really hits home—an "if this goes on" scenario—can't we still heed the warnings?

Are we at war with our planet? Or is our (not Gaian but instead possibly Medean) Earth somehow inherently programmed to destroy life every once in a while, geologically speaking? Do any species manage to get out of this alive? Recent data now suggest that even before the (unavoidable) six-mile diameter asteroid (or comet) impacted Earth 66 million years ago, the last dinosaurs were subjected to a hellish (endogenous) global warming epoch, thinning their ranks until *their* Doomsday Clock inevitably struck.

The metaphorical struggle between a powerful ape and a ferocious dino-monster of doom—which never happened in geological history—is evocative. Hence, this idiosyncratic tribute is, in part (if I may be so bold), an ode to an overtly silly, early, old 1960s Japanese monster movie featuring a symbolic confrontation between a giant ape and a world famous dino-monster—Toho's *King Kong vs. Godzilla*—yet a film utterly contrasting with messaging conveyed in the lead-up to Legendary's anticipated blockbuster, 2021's inevitable movie showdown *Godzilla vs. Kong*; monsters battling amid a Hollywood-engendered planetary crisis. Yes, heroic Kong! Exploited by man yet natural conqueror of both the fierce Tyrannosaurus and the Empire State Building—the early 20th century's technological wonder of the modern world. Versus titanic Godzilla, reluctant child of the hydrogen bomb, tragically irradiated in the fiery cauldron of mankind's overweening tendencies.[7] It just couldn't get any better than this!

The former Japanese film of 1962 lightheartedly characterized the two monsters' metaphorical battle in the face of a contemporary Cold War affliction; never as grimly with respect to the chilling prospect of nuclear Armageddon via the hydrogen bomb as in Toho's inaugural Godzilla performance—1954's *Gojira*. Now, a host of current 21st-century crises impact civilization broadly on so many interrelated fronts. Gaia, herself, wasn't conceivable until the mid–1960s, during early years of America's desperate space race. However, until Legendary threw their hat into the daikaiju filmmaking ring during the 2010s, the wave of global concerns reflecting on Gaia's role or regulatory influence remained generally neglected or delimited in the giant prehistoric monster movie arena. With acceleration of worsening planetary environmental conditions projected through coming decades, Legendary chose to repackage buildup toward Kong's inevitable rematch with Godzilla in 2021 (thematically, in what is arguably a paleontological Gaian context), reflecting current qualms and apprehension—our precarious sustainability. The deep past has returned to haunt and forewarn us.

I was born in the year of Godzilla's birth—1954. The first two movies I can recall seeing, both witnessed during the late 1950s, were Disney's *Sleeping Beauty* (1959) with its climactic fire-breathing dragon, at a packed theater, and RKO's *King Kong* (1933) on a television in our then Cambridge, Massachusetts, apartment. These movies left indelible, subliminal impressions upon me and so now—voila—here's a new book about a fire-breathing dragon and that famous super-ape.

Since adolescence I've been fascinated with geological theories of catastrophism and mass extinctions. Perhaps this is because, then as a third-grader, I so recall October 1962's Cuban Missile Crisis, then starkly accepting an impending doomsday. In retrospect, that crisis—coupled with prevalent nuclear themes in contemporary science fictional cinematic and television programming then, which I so strangely yet enthusiastically assimilated—shaped my global pessimism. (Being a lifelong Chicago Cubs fan certainly hasn't helped this dismal tendency.) I reminisced about that early coming-of-age experience in my Foreword to Mike Bogue's book *Apocalypse Then: American and Japanese Atomic Cinema, 1951–1967.*[8] Consequently, pessimism is partly the tone of this book as well.

At the time, dinosaurs were presumed to have gradually succumbed to their "inevitable" extinction, but then several cataclysmic theories entered popular culture, including the stunning prospect of nearby supernovae that may have irradiated Earth, sterilizing the planet, akin to a major nuclear holocaust although stemming from outer space. Then along came the 1980 asteroid theory of those unlucky dinosaurian extinctions that shifted focus of scientific investigations into mass extinctions of the fossil record. In each of my books I've incorporated chapters devoted to those fascinating mass-extinctions theories. Decades later, learning about these theories strikes me as such a nostalgic period in my life. And yet global concerns over environmentally triggered mass extinctions and pending catastrophe in the Anthropocene are more vibrant than ever. Extinction happens! Just surf the internet if you don't believe.

I first got into monsters during the very late 1950s and early 1960s. As mentioned, it started with *King Kong*, which I first watched with my dinosaur-loving dad in Cambridge, Massachusetts. Then after we moved to Park Forest, Illinois, in August 1961 there were a number of local horror programs that I grew accustomed to watching during later afternoons after school or late at night. By then I was smitten by issues of *Famous Monsters*, lovingly perused on dime store shelves while my parents went shopping. I yearned to see television broadcasts of those movie classics as emblazoned within hallowed monster mag pages—films starring Bela Lugosi, Boris Karloff and Lon Chaney. The very first of the Universal classics I recall watching on television was *House of Frankenstein*, which still airs periodically on Svengoolie's (Rich Koz's) hosted Chicagoland Saturday night show. In first grade I missed seeing *Gorgo*, but was able to see 1960's *The Lost World* at a downtown cinema with my grandparents, and then of course both *Godzilla, King of the Monsters* (1956) and *Gigantis, the Fire Monster* (1959) made their rounds on late afternoon Chicagoland television. Yes, by age eight, I had become a full-fledged paleontology, science fiction/horror and dino-monster-loving nerd!

Insensible wars raged during my junior high and high school years, but Captain Kirk and company offered hope for our future through reasoned and logical resolution time and again against seemingly insurmountable obstacles in *Star Trek* episodes. And while Godzilla defeated the symbolic Smog Monster the year after the first Earth Day (did Gaia care?), apart from television and cinematic screens, circumstances weren't really improving on a planetary scale. We also triggered environmental crises and a "population bomb." In fact, the downward spiral has only led to where we are today—a world irreversibly out of balance.

Despite repercussions of the Great Depression, a sense of hope for our future reverberated throughout as late as during the 1939 New York World's Fair.[9] However, in World War II's wake, the ensuing 1950s decade indelibly demarcated by the Soviet Union's development of the hydrogen bomb (aggressively responding to our own "Super" weapon) and later their successful launching of Sputnik turned tables on American optimism—soon replaced by pervading angst that has never fully dissipated.

In regard to the ecosphere, America's Environmental Protection Agency, for example, was a good start but look at where we are now half a century later. It is regrettable that my generation, overall, accomplished insufficiently in stemming the tide that so many scientists (and film producers!) forecasted decades ago. Increasingly vaunting ourselves somehow above and beyond a natural order, are humans failing as a natural species?

My adult interest in Gaia is tangential to my college and graduate school curricula, wherein I studied geochemistry of natural systems augmenting my chemistry major (yet related to my subsequent three-decade career in the environmental sciences). Academic encouragement and close instruction from an esteemed geochemist enlightened my way even though Gaia theory was never part of my formal studies. For many years, perhaps, our concept of a living Earth was relegated in popular culture to a 1970s television commercial for Chiffon Margarine featuring actress Dena Dietrich's humorous personification of Mother Nature, whose tagline was "It's not nice to fool Mother Nature."[10]

As time slipped by, increasingly, those dino-daikaiju and assorted dino-monsters of fantastic literature and film came to represent our plight. A doomsday monster of our own making had been unleashed that could only be defeated in science fiction tales! Meanwhile, during the environmental movement, Mother Nature, morphing into Gaia, reared her angry head. We own the reality of this eco-environmentally plagued situation. Hence Legendary's timely reboot of a quaint, old classic monster movie (and others leading up to it), although now reflective of contemporary, 21st-century issues plaguing the planet.

Godzilla and Kong weren't the first, or most famous, primordial monsters to storm pop-culture, or man's consciousness. During the early 19th century, "Frankensteinic" *Iguanodon* was foremost in terms of its presumed great size and powerful mien. It was a titanic animal that nobody had encountered before in imagination. Decades later, in the early 20th century, monstrous *Brontosaurus* seized human cogitations; then later, formidable *Tyrannosaurus rex*. But as we know, Rex was defeated by tragic, heroic Kong. And so, the question remains: could any giant monster prevail over him?

Now a few words concerning what this book is *not* about. This book is not a complete Godzilla filmography. Only a handful of Godzilla titles are discussed here (as well as several novels and graphic novels). Nor is it a compendium of all things dedicated to Godzilla or Kong. This is also not a dinosaur book, per se—although my premise is that the monsters featured within are anachronistically prehistoric or derived science fictional dino-daikaiju. Although this book is not an environmental science text, I invoke geological/paleontology ideologies and terminologies, especially those central to Gaia theory. However, because lately there has been a heightened public awareness concerning dangers posed by climate change, many readers may find my original take on relations between Gaia and daikaiju within interesting.

Here I must distinguish *real* dinosaurian ecological catastrophes of Earth history as known to science, such as that occurring 66 million years ago, portrayed by artists and writers likewise in fact and fiction, from science fictional apocalypses afflicting civilization portended by not-quite-so-real dinosaurs—dino-daikaiju—or other Kaiju of film and literature. While to 21st-century man both varieties would seem metaphorical in their distinctive ways, this book features the latter catastrophic category, not the former.[11]

I conceived the core idea—Gaia, Kong and Godzilla, interrelated via Legendary's productions—in August 2019 with early chapters fervently completed by that December. But meanwhile the key new movie I needed to assimilate, *Godzilla vs. Kong*, was postponed from an earlier March 2020 release date until March 2021!

I am not a movie critic, and so you will not find ratings for films discussed here. Nor shall I (over-)emphasize technically how these movies got made—cinematography, judging acting performances, and use of special effects and computerized programs (CGI). References within relevant chapters provide further information on such matters.

Also, this isn't intended as a "woke" cultural-awakening exercise. We, mostly, through the media, are already quite aware of what's happening to our world, environmentally speaking, regardless of our attitudes thereof. But isn't it further interesting to consider how, contemporarily, America's movie industry—relying on giant, anachronistically prehistoric, godlike monsters—projected our new-millennial planetary plight? After all, at its heart, any Godzilla movie is usually more than simply a movie about Godzilla rough-and-tumble battling formidable adversaries.

My primary purpose is to consider messaging, meaning and metaphor therein, even though readers may have strong opinions—pro or con—concerning each movie. While a featured film may not have been particularly popular or well regarded by fans or critics, this is not to say there still may have been significant or interesting ideas raised in the story, which remains the tenor of this book.

G-Fest XVIII (July 2011) was one of my favorites for a number of reasons. For present purposes, though, this was the event where (although I was a little tardy in attendance) I witnessed the lion's share of Sean Rhoads's presentation on scholarly aspects of Toho's *Godzilla vs. the Smog Monster*—a focal point of his master's thesis. I introduced myself to him via e-mail thereafter, while articles by him appeared eventually in *G-Fan* magazine and eventually a book (co-authored by Brooke McCorkle).[12] Stoked by his inspiring presentation, I began considering other original, tangential angles to pursue, ideas gestating over months and years: this book is one byproduct of that singular afternoon.

I also recall only a handful of years ago spying a blurb in *Time* (no less) forecasting a Warner Brothers film on the horizon for 2020, namely a remake of *King Kong vs. Godzilla*. This no doubt planted a distant seed, considering that no one had written a book focusing on these two iconic monsters. And yet even then I considered that such a volume could really be something ... more.

Allen A. Debus
Hanover Park, Illinois
July 2021

Introduction

Two Towering Titans of Terror and Tumult

For decades, to me, Kong and Godzilla just seemed to complete one another, even if they'd only battled in a single film—a non-decisive score (settled in 2021)! The primal theme of mammal-versus-reptile seems ingrained, "universal. Does it have anything to do with the Cretaceous-Paleogene mass extinction event 66-million years ago, when—as the story is usually told, dinosaurs lost in the great evolutionary sweepstakes, whereas we, or rather our mammalian ancestors, 'triumphed'?"[1] Therefore, shouldn't we naturally root for Kong (instead of Godzilla)?

By August 1961 I'd already seen *King Kong* with my dad at least twice, televised. What an epic thrill ride that movie was! He, in turn had seen Kong at the movies, escorted by his parents—the same year they saw animatronic prehistoric animal displays at the Chicago World's Fair: 1933. And so, if you asked me during the summer of 1961 which was my all-time favorite movie, the answer, in a heartbeat, would have been *King Kong*, featuring its heroic yet tragic dinosaur-thumping brute. But that fall, things changed, thanks to local Chicagoland television networks, when I became exposed to Japan's radioactive colossus—Godzilla!

By 1961 the world had changed considerably since 1933. World War II had been fought and won. Nuclear weapons had been invented and used both in war and repeatedly in test operations. The Soviet Union had quickly followed in pursuit of these bombs thanks to spy tactics. Korea had come and gone, but still not quite. While America's Kong was a relatively safe monster, an endangered crypto-creature beast confined to his insular lost world, not posing dire threats to the planet— Japan's towering Godzilla was another matter entirely, reflecting what

Man was beginning to fear most: an end to civilization and utter annihilation from the bomb!

It's difficult to recall exactly when I first saw *Godzilla, King of the Monsters* on a Chicagoland network, but Ted Okuda's and Mark Yurkiw's 2007 book *Chicago TV Horror Movie Shows: From Shock Theatre to Svengoolie* offers clues.[2] My experiences with Godzilla (and Gigantis) surely must have come from broadcasts on programs like *Thrillerama, The Big Show,* or *The Early Show,* all airing in Chicagoland during the early 1960s. When it came to monster movies then I was an equal opportunity black-and-white-screened TV-watcher. Godzilla and Gigantis were rare yet recurrent features during this period, and so I became enraptured with this formidable, fire-breathing, seemingly invulnerable "dino-dragon" plunging the planet in peril.

Toho's movies were dark, somber—unlike Kong's RKO debut. While 1933's Kong seemed heroic and tragic, Godzilla was angst incarnate, a dreadful omen. Yet, as I realized even then at so tender an age, the Japanese films reflected something sinister and disturbing, ineffably precarious and threatening about modern times. And yet how different were *those* times from 21st-century modernity!

Sometime in early June 1963, sight of a small ad in the Saturday movie section of the *Park Forest Star* shifted my world. Suddenly my mind was in a complete whirl. It was as if everything I knew about monsters and dinosaurs led to this very moment! There on the newspaper page I saw a mesmerizing drawing showing "The Two Mightiest Monsters of All Time ... in the most colossal conflict the screen has ever known! KING KONG VS. GODZILLA." Air force jets and helicopters were zooming in, train cars were falling from Godzilla's clutches, buildings were toppling and ships were torn asunder in frothy seas below their towering figures. In this ad Godzilla looked absolutely formidable, dinosaurian. But Kong ... well ... he didn't look right, not quite like I remembered him from his prior film appearance when he defeated the *Tyrannosaurus* and *Pteranodon*—only, bullet-riddled, to tumble off the Empire State Building. But maybe that was because Kong had only been ineffectually portrayed in a drawing, instead of a photo. I simply *had* to see this movie. ... ASAP!

And so, our dad—being an old-timey Kong lover—took my brother Rick and me to see the movie that very afternoon. I still remember the thrill of standing in line outside the theater as the crowd was slowly escorted inside, seeing those black-and-white movie stills posted outside the ticket booth, showing photos of scenes we were about to solemnly witness. Now Kong's appearance seemed even further, well, "off," but I couldn't wait to feast my eyes on Godzilla—*in color,* no less!

Nevertheless, our dad must have been snoozing in his seat by the time Kong also was being lulled to sleep with soma berry mist while sitting atop the Diet Building.

Well, as we know, the two monsters fought to a draw, but to me, fisticuffs aside, in retrospect, Godzilla won overall because Kong's suit just didn't seem Kong-ish or gorilla-like enough, while Godzilla had never looked better. Throughout the movie I kept forcing myself to foolishly believe that this particular *King Kong* version was the same as seen before (in the 1933 film) ... but convincing myself of this fact was as difficult as trying to understand how or why Kong had gotten so huge. (I was only eight and a half years old then, and quite willing to be gullible whenever necessary to suspend my own disbelief.)

A scene from *King Kong vs. Godzilla* oddly imprinted itself on my psyche. It stems from that part of the movie where Godzilla has destroyed a train and is pounding relentlessly toward a frightened, fleeing female. I had recurrent harrowing nightmares where I was playing in the (suburban) grassy field behind our house; then suddenly air raid sirens would begin blaring! Then, out of a rapidly expanding tornado storm cloud approaching from the north, Godzilla would materialize— heading, stomping directly toward (gulp!) me ... *me*—left with no place to hide or duck and cover. But remember, it was still Cold War times.

Themes of pending nuclear threat, and also lackluster explanations to describe Godzilla as a dinosaurian, were projected in the movie. Radioactivity is detected when a nuclear submarine is destroyed in an early scene while Godzilla emerges from a glacier—a nice tie-in to the prior film's icy conclusion in *Gigantis, the Fire Monster.* And until the end, when it is decided that the best way to vanquish both monsters— Kong and Godzilla—is to let them destroy each other on Mt. Fuji (e.g., because they're instinctive rivals), use of the atomic bomb remains a possible, last-resort alternative to relying on conventional weaponry. The atomic bomb is specifically referred to a handful of times throughout the movie. At one point it is mentioned that ten million people might perish if Godzilla reaches Tokyo—hence the necessity of pouring millions of Caltex gasoline into a deep trench (which of course would probably result in a major groundwater pollution problem), and amping up those electrical high-tension wires around the city (that Kong later uses to floss his teeth).

The prehistoric provenance for both monsters certainly seemed so exciting to my dinosaur-minded self then. Following Godzilla's dramatic emergence, American news commentator Eric Carter states how the world is stunned to know that prehistoric creatures are alive in the 20th century! A scientist then states that Godzilla is akin to a cross between a tyrannosaur and plated *Stegosaurus*, having thrived 97 to 125 million years ago, founded on Japanese fossils resembling the Godzilla species. Viewers are supposed to accept that its resurgence in modernity owes something to suspended animation. Regardless, I was enchanted that the scientists founded his argument on basis of a children's book[3] shown in the movie that we actually had at home. Meanwhile, Kong's prehistoric status was taken more for granted, since audiences were more familiar with his 1933 exertions on Skull Island. Nonetheless, despite their brain size disparity, this pairing of giant monsters were considered hated enemies, further suggesting prehistorical conflict between them.

I came away absolutely enthralled with what I'd witnessed on the big screen that afternoon. A year later, our maternal grandmother, Justina, purchased those 1964 Aurora model kits of both Godzilla and King Kong—whose size and appearance had evidently been based on the 1933 movie, fortunately not the '63 film. I immediately latched on to the Godzilla model, while my younger brother Rick kept Kong. Our toy train set would soon be ravaged by these creatures out of the prehistoric realm. Months later, the Godzilla board game arrived (although artwork on the box cover proved far more enchanting than the inside contents). When the Academy Awards were televised, I naively believed that *King Kong vs. Godzilla* might win for its special effects! I remember being so disappointed when the movie wasn't even mentioned (or nominated).

I had always thought that Kong was supposed to be the good guy in this movie, and Godzilla the scarier "bad" monster. Partly this is because Kong, as an anthropoid, was more akin to our species, while Godzilla was a saurian dino-monster, emitter of radiation. But surprisingly a 2019 *G-Fan* article[4] altered my view on Godzilla's motives during that mid–1960s period. Maybe Godzilla was just minding his own business all along—as Curly Howard of the Three Stooges might say, "I'm just a victim of circumstance."

A few years later, from my then-11-year-old perspective, with the Cuban Missile Crisis safely behind us, and given that the Soviets hadn't started a nuclear war (as some feared) in the wake of JFK's assassination, for a little while the angst of living in the nuclear age diminished slightly. Godzilla no longer seemed to be the looming metaphorical radioactive threat it had once posed to the world as projected, say, on Chicago's *Thrillerama* show. For a discrete interval of time, it didn't seem as if self-extinction of the human race was imminent or plausible, even though Toho conjured up another world threat for us to face—Ghidrah (later known to American audiences and fans as King Ghidorah)!

Yes, in the fall of 1965, the *Park Forest Star* seduced me once again, transmitting with such urgency and with evidently so much at stake: "Ghidrah! The Three Headed Monster Battles Godzilla, Mothra and Rodan for the World! All New Sights! Never to Be Forgotten!" It all looked so cool.

No Kong this time, but Rodan was a welcome sight—a television favorite by then and from newspaper advertising this new monster Ghidrah (aka Ghidorah) looked eminently cool. (I just couldn't wrap my head around how fuzzy-looking Mothra could fare well against these bigger and more formidable brutes—but then that's why I needed to see the movie.) Now Ghidrah represented an all-consuming world threat, supplanting Godzilla's prior persona, role and meaning—now if only Earth's terrible monsters could cooperate to defeat the even more menacing outer space destroyer!

Crashing in from space not unlike that asteroid which doomed the dinosaurs 66 million years prior,[5] Ghidrah took fiery form in a dramatic scene! Meanwhile Godzilla and Rodan were already bickering, before in the end teaming with a sensible Mothra larva to fight highly impressive Ghidrah, lest our planet be utterly destroyed by this common enemy. Yes—there was that geopolitical lesson therein. Stemming from an inexplicable heat wave gripping the island, a sense of utter urgency portending that "something terrible or cataclysmic was going to happen" pervaded throughout the first half of the film—thanks to a Martian Prophetess. She warns of the arrival of giant doomsday monsters and the ensuing destruction to come. Humans would suffer extinction when Ghidrah passes, an outcome which already happened to the civilization on Mars—now a "dead world." She even prays to the almighty universe to save our planet from annihilation! One might say she's ... my favorite Martian. Yet when scientists hypnotize her, probing for the truth during that "You're-in-a-deep-sleep" scene, our dad was probably already snoozing again too.

Meanwhile Japan's authorities claim they cannot authorize use of atomic weapons ... especially when Mothra—the smartest and bravest of the monsters—can convince Rodan and Godzilla to collectively fight Ghidrah because with everything in the balance, Earth still belongs to them as much as it does to us.

While the buildup to the climactic battle was quite good, the monsters' comical antics ultimately seemed undignified, undermining the gravity of circumstances Planet Earth was supposedly facing according to the Prophetess. Thereafter, Godzilla's persona transformed into a saurian super-hero, while Kong's diminished (until the new millennium).

In 2019's blockbuster entry, Legendary's *Godzilla: King of the Monsters*, its title recalling that of the very first Godzilla movie I ever saw, although apparently founded on the giant monster battle I recalled in 1965 as *Ghidrah the Three-headed Monster*, a lead character, Emma, states: "The mass extinction we have feared has already begun, and we are the cause ... we are the infection. But like all living organisms, the Earth unleashed a fever to fight this infection" represented by the mythological-sounding "Titans" which will "guarantee that life will carry on." This theme and perspective is rather Gaian (e.g., treating Earth in totality as a living organism, as theorized by James E. Lovelock, as for example outlined in his popular 1979 publication, *Gaia: A New Look at Life on Earth*). But for the view of humans as infection, or perhaps a virus, I recollect Loren Eiseley's views, expressed in his 1970 book *The Invisible Pyramid*.[6]

Clearly the concept of extinction as conveyed in a 21st-century style seems more spectacularly dire, as opposed to how us monster-land kids received analogous messages six decades ago. This must mean planetary circumstances truly are worse now than during days of my youth.

How Kong's reemergence in 2021—as tragic hero or Godzilla's subversive foil—might soften Mankind's ultimate fate remained to be seen. Which monster would likely become Earth's mightiest savior, Godzilla or King Kong?

CHAPTER ONE

King Kong and Godzilla

*Immortal Adversaries—
21st-Century Daikaiju*

As sage Carl Sagan (1934–1996) once speculated, "In the ... night-time stirring of the dream dragons, we may each of us be replaying the hundred-million-year-old warfare between the reptiles and the mammals."[1] But from conflict and chaos comes stability. Natural systems equilibrate, seeking balance. And foremost of all monster-dom, perhaps the most sensational, celebrated and (allegorically) urgent cinematic projection of how planetary imbalance and ecological distortion caused by Man's ruinous activities may be restored is conveyed through King Kong's and Godzilla's tensional, anthropomorphic interplay—that is, gargantuan mammal pitted alongside, juxtaposed with titanic dragodinosauroid imbued with quasi-human faculties.[2] And yet here I do proclaim Kong *and* Godzilla—not "versus"—for they stand together, lest we succumb to Gaia's—once conceived as the most gigantic organism known—ravages.

What on Earth? Yes, exactly—that's what this is all about.

Pivotal, prominent themes in the dino-monster Legendary new millennial film series, emphasizing a "Monsterverse," include essentiality of ecosystem balance, mad Frankensteinian technology, what is natural versus (from our exalted perspective) what is not, and the prospect of apocalyptical human extinction—coming full circle. Also apparent throughout, largely unstated, is influence of Gaian Earth Systems Science.[3] Is Nature, as Man has long felt, something to be subdued and conquered, or an entity that cannot be harnessed and, in fact, *controlled*? Arguably such ideologies therein—featuring distinctively American kaiju—are more elaborately plot-interwoven than as represented in Toho's original thread of daikaiju films spanning over six decades. Why this intricately, invigorated emphasis in Legendary's 2010s giant dino-monster film canon? Perhaps spurred by deepening

Figure 1: Small advertisement appearing in *Boys' Magazine* (Oct. 28, 1933) for John Hunter's serialized 1933–1934 novella *Menace of the Monsters*. The mystique of a Kong-type creature battling a terrifying carnivorous dinosaur was at the forefront of fantastic literature—thanks to RKO's release of *King Kong* earlier in 1933.

worries over accelerating climate change, is Gaia, recognized as "one of the crucial ideas of the twentieth century,"[4] a western phenomenon, herself an answer?

Readers may be unfamiliar with the brilliant Gaian (e.g., "living biosphere") concept, which by the 1980s had entered the realm of popular culture. Some may equate with that old nurturing concept of Mother Earth, but that is far too embracing and anthropomorphic. Accordingly, here is a concise summary (written by historian of science Peter S. Alagona in 2010):

> The Gaia theory conceptualizes Earth as a single system, with *negative feedbacks* that maintain stability under normal conditions and positive feedbacks that can vault the system into a new state under exceptional circumstances. Gaia has had a contested history. The concept was not

entirely new when (James) Lovelock began to develop it in the 1970s, but his formulation was novel, and it elicited passionate refutations and endorsements. Many scientists regarded the theory as New Age bunk or, worse, ignored it entirely. But this changed during the 1990s when studies of global warming catalyzed the formation of a new integrative field, called Earth system science, and established Lovelock as one of its foundational thinkers. ... Lovelock experienced decades of marginalization, and even ridicule, before finally receiving recognition at the beginning of the 21st century.[5] (my italics)

Furthermore, as American paleontologist Peter Ward noted, Gaia is at the heart of (western) modern environmentalism. Throughout, British scientist James Lovelock's (b. 1919) considerations and innovative thinking have been instrumental toward advancing this profound scientific and societal movement. As we shall later see, however, there is a dark counterproposal to Gaia, namely the "Medea hypothesis" (as coined by Ward) contending that "the evolution of life triggered a series of disasters that are inimical to life and will continue to do so in the future." Without Gaian philosophy, a Medean hypothesis wouldn't have resulted.[6]

An important distinction between the two ideologies concerns Man's role in nature. Self-regulating Gaia would uncaringly, unconsciously, non-purposefully (via biota and biogeochemical feedback processes) replace us—polluting mankind—if we trend toward altering planetary environmental conditions necessary for life's continuity through future geological time. Gaia always looks out for herself. In contrast, according to the Medea hypothesis, "Not only are we the *major success*, we are the only hope for life to save itself from itself"[7] (my italics). Although prospect of our pending extinction forms an alarmist core of the Legendary dino-monster filmography, "Mass extinctions play no part in the Gaia hypotheses. ... Gaians consider them 'Gaia neutral'..."[8] However, they are prominent in Medean geohistory.

Although without transparency, Lovelock's better known core ideas (as well as aspects of Ward's) are interspersed throughout the Monsterverse, and allied dino-daikaiju films in Legendary's 21st-century framework through the concept of restoring natural balance. To my knowledge, this particular suite of films has not previously been discussed in a Gaian/Medean context addressing implications for American-made, filmic dino-daikaiju thereof.

Death of Godzilla and Shifting Perspectives

As the new millennium beckoned, Toho's Godzilla franchise nearly died twice. First, Godzilla was killed in 1995's *Godzilla vs.*

Destoroyah—thus ending the (second) "Heisei" series—only to return in an ensuing "Millennial" series. Following 2004's *Godzilla: Final Wars*, the plan was to scrap, refresh and retool, thus capping Godzilla filming for (over) a decade. Meanwhile, efforts to create a new comparable and distinctive, definitive American daikaiju failed. 1998's version of *Godzilla*—TriStar's mutation—did not appeal to scores of older generation G-fans, while Fox Network's televised 1998 rival, *Gargantua*, is forgotten. Peter Jackson resurrected iconic Kong in 2005's updated remake, *King Kong*, fortified with dazzling computer graphic effects, and 2008's *Cloverfield*, a giant monster exceedingly difficult to visualize in the story's "recovered" footage, evoked unsettling terrorism akin to 9/11.

None of these entries paved the way for the more successful series that would soon come. And none—each relying on familiar, classic movie monster themes—introduced Gaian mindfulness to audiences either.

When Toho finally returned with their well-received 2017 entry, *Shin Godzilla*—their titular monster now cast as a rapidly evolving product of radioactive pollution—however, an assortment of symbolic, Legendary American daikaiju and paleo-monsters had already paraded through our cinemas during the 2010s, reflecting ecological and environmental concerns of discernable western flavor.

Certainly, Toho and other producers of Japanese daikaiju films more than just dabbled in environmental themes at the core of the giant monster menace. 1971's *Godzilla vs. the Smog Monster* was their flagship, yet not the only such example. In their book *Japan's Green Monsters: Environmental Commentary in Kaiju Cinema* Sean Rhoads and Brooke McCorkle detailed such concerns as explored throughout a host of Japanese science fiction films.[9] Japanese concerns over many forms of pollution often exceeded our own, politically, as then reflected in many 20th-century contemporary filmic portrayals, including radioactivity as well as climate change, and are prominently reflected in many daikaiju films. However, western influences of Lovelock's Gaia—allusions to which are absent in Japanese filmography—are recognizable, albeit indirectly, in Legendary's Monsterverse filmography. Is this a good sign? Despite the politics of our anxious times, are we, the western populace, the masses, now finally just getting it?

For many years in movies (and sci-fi literature), giant monsters (often dinosaurian daikaiju in sci-fi provenance suggesting disequilibrium through stark juxtaposition of the prehistoric suddenly come to life in modernity—more or less, analogously, like antimatter co-existing with normal matter) threatened mankind by attacking and/or demolishing major metropolitan centers. Max Page's 2008 book *The City's*

End: Two Centuries of Fantasies, Fears, and Premonitions of New York's Destruction outlines how New York City has stood metaphorically for civilization teetering on the brink of apocalypse.[10] We are helplessly drawn to witness the fantasy of our total destruction through artful means, a kind of "disaster porn."[11] And while Page's expert analysis addresses horrors of looming climate change, citing films and art inspired in part by Al Gore's *An Inconvenient Truth*,[12] it doesn't consider a more enveloping, universal form of planetary devastation—that could happen if unminding Gaia itself or herself, sensing human-caused imbalance, strikes at the root cause. Us.

An Epoch of Extinction

Humans have so transformed the face of Earth that geologists have carved an epoch for ourselves in the industrial age—the Anthropocene, geochemically recorded in developing strata dating to the first radioactivity-spewing atomic bomb explosions of 1945 (or possibly stretching as far back to the beginning of the Industrial Age during the late 18th century). We are in the midst of a Sixth Extinction, according to paleobiologists, caused by human waste and climate change threatening not only other marine and terrestrial species, but also ourselves. Earth's ocean-atmosphere system is rapidly being altered, and not to our advantage. If Amazonian Rain Forests—the "lungs" of our planet, generating up to 20 percent of our oxygen—disappeared, burning at unprecedented rates in 2019, would we all inevitably suffocate? Would the seas become anoxic, "Strangelove" oceans occasionally stirred up by category 6 or 7 hurricanes? Theoretically, yes.

Warner Brothers and Legendary Pictures reflected angst many of us might have been increasingly sensing by then with their 2013 movie *Pacific Rim* (directed by Guillermo del Toro), and a tie-in media novel written by Alex Irvine. While the Japanese word "kaiju" had been familiar to G-fans for decades, *Pacific Rim* certainly popularized this term in the western mainstream. Is our own extinction on the horizon, in the grip of metaphorical kaiju that ultimately are of dinosaurian origin, storming from Hell, the bowels of our planet—arguably in a Gaian fashion? Daikaiju would appear to be awful products of the Anthropocene, harbingers of Doomsday.

Although imbued with DNA, as clarified in Irvine's novelization, "built on a template of silicon," those *Pacific Rim* kaiju are not only "like a dinosaur, kind of, only an order of magnitude larger than any dinosaur that ever lived,"[13] but in fact they're bioengineered evolutionary

Figure 2: Poster advertisement for Kaiju-infested film *Pacific Rim* (2013).

derivatives of Earth's Mesozoic Era dinosaurs—such as the more familiar genus *Stegosaurus*! Framed in metaphor, this is man (e.g., mammals) versus (pseudo-dinosaurian) reptiles once more.

Their presence is a warning to us, polluting humans, who are optimizing, "terraforming" the planet, into a world increasingly afflicted with critical factors leading to mass extinctions—acidic oceans and depleted atmospheric ozone—perfected for kaiju habitation and domination. Mankind is losing the Kaiju War!

Telepathic kaiju erupt through a dimensional portal known as the "Breach" in the Pacific Ocean abyss, and are assigned category ratings of from 1 through 5—the worst level of highest "ferocity." Thus, hurricane category lingo is applied to each kaiju as they approach shorelines, plaguing mankind. That through time kaiju increasingly trend (evolve) toward higher category ratings and larger sizes—at the story's climax, kaiju "Slattern" is 600 feet long—suggests our ever-worsening human-caused climate change crisis.

Kaiju wage an evolutionary arms race with humans. For, as stated in Greg Keyes's 2018 official movie prequel, *Pacific Rim: Uprising*, "You never knew what the Kaiju were going to be like, especially … when they seemed to be adapting to the tactics and technology of their human adversaries. What worked five years ago … might not work now … once again innovation would be important, much as it was in biological evolution."[14]

In our worst nightmares, such (to us, alien) kaiju would appear as manifestations of Gaia's revenge, because in a post-kaiju world, life would continue on Earth—although without an unnecessary species, *Homo sapiens*. Armed with robo-tech Jaegers, humans score key victories against kaiju. There is a happy Hollywood ending. The strategy to stem the tide of kaiju invasions hinges on sealing the Breach from their mysterious place of origin, the "Anteverse," ironically relying on our destructive capability, a thermonuclear weapon. However, due to harrowing circumstances faced by our heroes, audiences must wonder whether "Precursors," who spawned kaiju in the Anteverse and who have been plotting kaiju invasions for a hundred million years, would eventually prevail in the long run.

In *Pacific Rim*, kaiju could be construed as Earth's negative, controlling feedback mechanism, restoring balance in retaliation to ourselves. However, analogous to Man's tendencies, kaiju instead may merely represent another positive, out-of-control feedback loop, leading to planet sterility. Which is the worst threat—Man or Kaiju? Or are we essentially the same, flip sides of the coin, depending on perspective?

The Precursors' earlier (Mesozoic) kaiju invasion was thwarted because Earth's atmosphere wasn't sufficiently contaminated yet. So ancestral kaiju, dinosaurs known to paleontology such as *Stegosaurus*, were forced to bide their time, awaiting our ascendance to do the atmosphere-altering job for them. (There is no mention of the asteroid impact theory that doomed the dinosaurs 66 million years ago in *Pacific Rim's* scenario.) Meanwhile, Gaia would appear not necessarily impotent, although presiding, uncaring of Man's continuation versus an eventual Precursor takeover in a polluted world optimized for kaiju existence. And so the ultimate warning becomes what happens to us— replacement—if we do not take care of our planet.

Precursors have destroyed many worlds throughout the cosmos before, storming from the Anteverse into our universe. One wonders whether in the future following Man's extinction Gaia would in turn usurp kaiju prior to their rendering Earth into another uninhabitable satellite.

In Greg Keyes' 2018 *Pacific Rim: Uprising* tie-in novel, we learn that a doomsday bomb, fused with explosive kaiju blood, has been devised to open a breach in the ocean floor that threatens to "trigger a catastrophe so large it would re-form the world to Precursor specifications. ... We could expect extinction of animal and plant life on an unprecedented scale. It might ... be greater than the mass extinction at the end of the Permian period ... probably (also) caused by volcanism."[15]

Although not a Legendary picture, dinosaurian connections

Figure 3: *Pacific Rim's* Leatherback was another Kaiju genetically derived from dinosaurian stock, as more fully elaborated in Alex Irvine's novelization.

are also evident in 2014's *Transformers: Age of Extinction* (Hasbro, Di Bonaventure, directed by Michael Bay). Several of the living robotic, alien Transformers even resemble *Pacific Rim's* kaiju dino-derivatives, featuring a fire-breathing pseudo-*T. rex* and a two-headed pterodactyl. Here, following an invasion of Earth and after planting metallic "seeds" 66 million years ago, thus ending the Mesozoic Era, today, the geologically excavated seeds made of "transformium" threaten extinction of Man as well—thereby linking Man's looming demise to the dinosaurs' sudden end.

Man Creates Kaiju

Anthropogenic impacts on climate and ecology are said to be superimposed on natural processes. But then aren't human influences also natural? In examining Man's fate in light of Gaian controlling systems, let's consider another set of Legendary's symbolic dino-monsters—genetic menaces posed in a laboratory-reconstituted *Jurassic World* (directed by Colin Trevorrow).

Mad Frankensteinian science lies at science fiction's core—dating back to Mary Shelley's (1797–1851) *Frankenstein, or the Modern Prometheus* in 1818. Michael Crichton relied on such forbidden technology in two fantastic novels of the 1990s in which genetic power, as wielded by Man, was presumed more potent than nuclear power in its destructive potential.[16] Later, in Sam Enthoven's 2008 novel *TIM: Defender of the Earth*, scientists create the world's first giant, genetically engineered dino-daikaiju from *Tyrannosaur* DNA. "What we've done ... is directly manipulate DNA ... to create something entirely new: a creature *based* on living, or once-living, things but that is in fact *entirely different*."[17] Certainly, once the proverbial genie is lured from the bottle, it cannot be reinserted (either in technology or in sci-fi literature)!

Hence, in Legendary's *Jurassic World* series (Universal and Amblin Entertainment, in association with Legendary Pictures, 2015), filmic extrapolations of Crichton's imagined *Jurassic Park* universe, cloning dinosaurs takes an even darker turn, when Man-the-creator, wielder of industrial fervor and fire, bioengineers entirely new forms of living dino-monsters considerably differentiated from those which previously existed during the Mesozoic Era and thus would seem far less "natural" than the original clan of supposedly real dinosaurs, revivified in Crichton's novels from DNA traces found in insect-containing amber. A strange unnatural history now unfolds. Or does it?

There is dichotomy between what most people regard as natural versus unnatural. If humans caused something or altered planetary processes in some way, such as by introducing new species, then usually those events would be considered unnatural, even though Man is also a natural product of evolution and merely another of Earth's fateful species. However, pollution on a planetary scale caused by any species, such as microbials that contaminated Earth's embryonic atmosphere with a persisting supply of oxygen during the Precambrian, may also be thought of as natural because that is how our planet has evolved through eons. Through whatever mechanism, life nested here 4.5 billion years ago, securing its way through the ages, prompting intricate Gaian responses to maintain a biogeochemical balance necessary for interactive ecosystems, various evolving forms of living things—to prevent life's extermination by maintaining more or less constant conditions through time (via homeostasis), despite perturbations.

Five major mass extinction episodes are recorded in Earth's geological history, including the end-Cretaceous event of 66 million years ago, leading toward our current wave of ecological devastation that has been dubbed the "Sixth Extinction," resulting from what is often perceived as an "unnatural history."[18] Lovelock would counter that while ongoing pollution rates—which must be curbed for survival—pose dire consequences for us and that of many other species, our actions are indeed natural, for we are living creatures.[19] Therefore, Lovelock contends, continental or even global scale pollution—think of it simply as organisms modifying their environment—must be a natural outcome in the Gaian system.

Theoretically and in our darkest exaggerated fears, it might be possible for Man's actions to overwhelm planetary restorative properties, to an extent of causing extinction of all life, although Lovelock dismisses the possibility. Humans simply lack capability to destroy *all* life on Earth, even as we continue to spew synthetic chemicals, radiation and other forms of pollution over land and seas … or conceptually wield that awesome genetic power in most unscrupulous ways imaginable, or yet not imagined. Or, say, even if Crichton's imagined reintroduction of dinosaurs into modernity gets completely, irrevocably out of hand.

In 1979 Lovelock posed a provoking thought experiment on how introduction of a scientist's genetically modified algal organism into nature could bloom in an out-of-control fashion leading to a sterile planet. "Its capacity to gather phosphorous rendered the environment barren for everything else … stricken Earth would move slowly but inexorably towards a barren steady state, although the time scale might be of the order of a million years …"[20] Yet bioengineered dinosaurs are much sexier to ponder than world-ending microbials (mere, yet irrepressible bugs and viruses), and so in blockbuster movies it is the former clan usually relied on to suggest a dawning genetic apocalypse, naturally engineered by Man.

Imagine a world in which "no one is impressed with dinosaurs anymore,"[21] due to their commonality in theme parks such as a "Jurassic World" on Isla Nublar. Then to attract crowds and gain profits, bioengineers must increase the wow factor, by inventing pseudo-dinosaurs, dino-monsters such as *Indominus rex!*[22] While nightmarish gate attractions may seem thrilling, in reality unleashing the genetic "genie" causes Jurassic World's disruption (e.g., containment breaches) on a level equivalent to a nuclear plant meltdown. (Think Chernobyl!) A flock of vicious, killer pterodactyls stream through a gap in a domed aviary. Ultimately, the presumptive remedy is fundamentally flawed. In *Jurassic World* human-devised controls are not reliable. Our safety is an illusion. After all, we are expendable.

Indominus may only be part dinosaur, but it also qualifies as kaiju. It kills for sport. It is clever, highly intelligent, capable of communication with related species (i.e., the cloned *Velociraptors*, which share a component of its programmed DNA). It outmaneuvers trained soldiers (i.e., the fully locked and loaded "asset containment team") with ease. When the quartet of weaponized "raptors" are called into action

Figure 4: ***Jurassic World*'s genetic Kaiju-esque aberration, Indominus rex (illustration by Mike Fredericks [2021], used with permission).**

to stop *Indominus's* advances, their larger genetic cousin stages a coup d'état: raptors betray their supposed human handler. And in the end, the less natural *Indominus* nearly defeats a more genetically correct *Tyrannosaurus*—yes, the original, aged tyrant king from the *Jurassic Park* story—in a bloody battle. During this carnage, in several shots the name Pandora can be seen on lit signage—underscoring what Man has done this time! Fortunately, *Indominus* is the only one of Dr. Wu's genetic monstrosities ("side projects") so far manifested upon Jurassic World.

To what extent would any of these test tube dino-species breed and proliferate in the wild, beyond isolated Isla Nublar, subjugating us? Crichton barely scraped the surface of this possibility in his 1990 *Jurassic Park* novel, while the idea became fully realized in 2018's *Jurassic World: Fallen Kingdom* (Universal, Amblin Entertainment, in association with Legendary Pictures/Perfect World Pictures, directed by J.A. Bayona).

Here, remaining exploited dinosaurs are considered by some as an endangered species, more so because a volcano on Isla Nublar is ready to blow, thus potentiating their (second) extinction via an "act of God." But a wealthy individual, Mr. Lockwood, former associate of John Hammond—creator of the original Jurassic Park—facilitates dinosaur protectionists' operations by rescuing several species during the fateful eruption. (Appropriately, there's a portrait of author Mary Shelley, Frankenstein's conjurer, visible on a wall outside Lockwood's private paleontology museum on his spacious estate.) Turns out, though, unknown to Lockwood, an ulterior motive for capturing dinosaurs from Isla Nublar, including the last remaining *Velociraptor*, is so that they may be auctioned at ultra-high prices, thus financing Dr. Wu's laboratory efforts using the latter's DNA to create a perfectly weaponized, newly designed species, a genetically enhanced Indoraptor (named from *Indo*-minus + Veloci-*raptor* = *Indoraptor*). This next-generation near-kaiju dino-monster would be a gene-coded blend of sampled *Indominus* DNA merged with that from *Jurassic World's* last *Velociraptor*, considered the second most intelligent species on Earth, augmenting an existing *Indoraptor* prototype—which gets loose before finally being outsmarted.

In the end, "life finds a way"[23] once more, though, as Lockwood's daughter Maisie—herself a bioengineered clone of his original (deceased) daughter, thus exemplifying inherent dangers of biotech—symbolically unleashes the captured dinosaurs into the North American wild, simply because, like her, they're alive. Here, the potentiality of proliferation is thematically evident, be it via nuclear weaponry or as

Figure 5: In the *Jurassic World* dino-monsterverse, Tyrannosaurus and Velociraptor (shown in this movie advertisement poster) play significant, albeit aberrant genetic roles in facilitating harvesting of *Jurassic World*'s Indominus and *Jurassic World: Fallen Kingdom*'s Indoraptor (author's collection).

underscored through analogous misuse of genetic power—knowledge and technology that is now fatefully out there for anyone to harness.

So is it under Gaia's or Medea's reign in which the mammal versus reptile war is waged, given that, due to Man's mastery of biotech, dinosaurs threaten to outstrip human civilization in a dawning Neo-Jurassic Age? And when do natural systems (or possibly those influenced from external cosmic sources), overwhelm Gaian controlling, restorative-balancing mechanisms? Is Gaia always resilient in exerting those negative feedback controls, neutral to biological systems, or vengefully self-destructive to certain organisms via harmful positive feedbacks, beyond an irrevocable tipping point?

"I am become Death"

In Warner Brothers' and Legendary's 2014 film *Godzilla* (directed by Gareth Edwards), it would seem civilization may survive after all because according to Dr. Serizawa's analysis, for uncanny reasons, Godzilla's presence in modernity is to "restore balance." Godzilla is represented as an anthropomorphized Gaian warrior manifestation seeking to maintain natural balance ... for our sake, to prevent replacement of our species.

Here, Godzilla is projected as a primordial creature stemming from a past geological era of elevated radioactivity. Likewise, his ancient nemeses, the parasitic bug-like, radiation-seeking MUTOs, evolved in that time. This would seem an intriguing, disbelief-suspending premise rather founded in fact, anticipated by insights of Lovelock and, independently, kaiju writer D.G. Valdron in his 2005 short story "Fossils."[24]

As Lovelock stated, four and a half billion years ago Earth's natural radiation levels were elevated relative to today: "The proportion of radioactive uranium-235 present in all the uranium on Earth is 0.72 percent. But when the Earth formed ... the proportion was 15 percent, and other radioactive material was also present in richer concentrations then than it is now." And yet Lovelock views the prospect of natural radiation as "fearful though it may be to individual humans is to Gaia a minor affair."[25]

Valdron's fascinating story introduced a gigantic Godzillean creature known as "Kaiju A," which plunders Earth. This dino-monster and others of its prehistoric ilk roaming the planet—evolved while exposed to concentrated radioactivity—were reawakened by our nuclear bomb detonations. According to Valdron, "Every seventeen million years the comets come to wake the Kaiju, who sterilize the world for the next cycle of life. We've woken them early; that's why they are confused."[26]

In the 2014 film, as well as Greg Cox's engaging novelization from the script by Max Borenstein, with lethal levels of radiation at the core of our problem Godzilla seeks to restore a semblance of natural balance while the MUTOs (Massive Unidentified Terrestrial Organisms) pose a more deadly threat to civilization—a fate shared by both Man and Godzilla. Balance can only be achieved this time if Man cooperates with Monster! Man's mining operations and industry—thus disinterring larval pod relics out of the past—are entirely natural. So if we inadvertently resurrect MUTOs from Time's tomb, then is Godzilla's revivification (during our Anthropocene Era) not just a harbinger of what may befall us, but also another example of an unconsciously exerted Gaian ("geo-physiological" per Lovelock) negative feedback control?

Scientists recognize a spectrum of Gaian hypotheses, ranging from the "strong" or "optimizing" Gaia, to those exerting lesser influence and control over Earth's climatic, biogeochemical cycling and ocean/atmosphere systems throughout geological time. Which versions are reflected in Legendary's *Godzilla* and its filmic series? And if none of the Gaian ideologies best fits the picture, then is Legendary's perspective instead Medean? Because answers to such arcane questions aren't fully apparent through movie analysis alone, we must consult Cox's 2014 informative novelization from the script to glean further insights.

This story begins with a symbolic pairing of containment breaches: radiation fuels the core problem faced by humanity. The first is a biological breach—mining operations in the Philippines expose a fossilized Godzillean skeleton permitting a MUTO male larva to hatch from a large spore, which escapes to a radioactive source in Japan, the Janjira nuclear power plant. Its bio-acoustical transmissions nearly cause catastrophic meltdown at the plant. A second dormant spore discovered in the fossil—containing a female larva, radioactive—is shipped to the Yucca Mountain Nuclear Waste Depository near Las Vegas, where it slowly revives, breaching that facility's confines fifteen years later. In one passage of the novelization, when informed of the female MUTO's advance toward Las Vegas, a skeptic claims this bit of fake news must be "a hoax like global warming."

Another containment breach occurs fifteen years later as well when, following a series of jolting electromagnetic pulses (EMPs) emanating from the cocoon of the larva which escaped to Japan from the godzillean fossil, a (winged) praying mantis–like, obsidian-shelled male MUTO emerges. (In its maturity the MUTO's EMP signal is sufficient to quash Godzilla's brilliant radi-ray in battle.) With this MUTO's metamorphosis completed—long fed by radiation remaining in the buried nuclear core of the destroyed Janjira power plant—the male hatches in a dramatic display of destruction. (Project Monarch scientists had been studying its bio-acoustically active cocoon, perceived as a "living fuel cell," in a revolutionary effort to solve the world's energy crisis.)

After securing a 300-kiloton thermonuclear warhead for their radioactive nest—likened to the fiery pit of Hell—the two bugs spawn in San Francisco, thus threatening the planet with an insurmountable population of giant invincible monsters. The breeding bugs may be likened to a nuclear "breeder reactor" potentially capable of usurping the planet. In this universe, man is insignificant, ironically a tiny insect by comparison, merely collateral damage. Meanwhile, the (extant) crypto-Godzilla (which we attempted to destroy in 1954 using nukes) reemerges as a tidal wave—an invincible natural force of potential cataclysm. (Later, Godzilla's emergence from San Francisco Bay is perceived as a "newborn volcano.") Sensing looming primordial danger and with human aid, Godzilla seeks to eradicate the MUTOs. The younger Dr. Serizawa's symbolic wristwatch of doom, a relic bequeathed from his father who created the oxygen destroyer, remains frozen at 8:15 a.m., Hiroshima time, August 6, 1945!

Dr. Serizawa (Ken Watanabe), cognizant of Man's arrogance, utters prophetically, "Nature has an order. ... A power to rebalance. ... I believe that he (i.e., Godzilla) is that power."[27] It's almost as if living Gaia deliberately summoned Godzilla on cue for the purposes of maintaining the human-civilized status quo. Or maybe, channeling Peter Ward's view, by adventitiously defending our continued destructive tendencies, instead, Godzilla—in thwarting the MUTOs—represents a Medean force indirectly far more effective in exterminating greater swaths of life than would be the case of a MUTOs takeover (i.e., given that if, instead, humans survive this threat, we can exact far greater damage to the Earth system than would the MUTOs).

Dare we meddling humans destroy Godzilla with an H-bomb (along with the MUTOs) at sea, lest we permanently disrupt Serizawa's perception of natural order? Or do we "let them (i.e., the three monsters) fight"? The original plan is thwarted when the male MUTO absconds with the warhead. Now if the nuke detonates within the city, radiation fallout would catalyze hundreds of eggs in the nest. Therefore, their ensuing fight, while efforts are made to defuse the bomb, becomes a Darwinian struggle, "survival of the fittest—on a grandiose scale."[28]

Despite Man's survival, perhaps a questionable outcome for Earth's overall geophysiology, there is another happy Hollywood ending: "Nature was at peace." With the world in balance again, Serizawa contemplates, "Godzilla destroyed the MUTOs ... as Nature *intended*,"[29] implying purposeful intervention. As Ford Brody notes, "Godzilla had done his part, ridding the world of the MUTOs." Faith in Nature is thus equivalent to believing in good of a newly identified natural force incarnate, heroic mega-saurian Godzilla, only appearing when the stakes are highest.

Medea vs. Gaia

We've adopted a more or less Gaian view in examining these several films, in which Legendary played a pivotal role. Lovelock understood the dark side of Gaian change. "Looked at from the time scale of our own brief lives, environmental change must seem haphazard, even *malign*. ... during ... abrupt changes the resident species suffered catastrophe whose scale was such as to make a total nuclear war seem, by comparison, as trivial as a summer breeze is to a hurricane."[30] Ward echoes the malignity issue, offering an even more sobering view. "Life itself, because it is inherently Darwinian, is biocidal, suicidal, and creates a series of positive feedbacks to Earth systems (such as global temperature and atmospheric carbon dioxide and methane content) that harm later generations. Thus, it is that life will cause the end of itself..."[31]

As Dr. Serizawa professes, "The arrogance of Man is thinking Nature is in our control, and not the other way around."[32] Lovelock stated, "It is not much comfort to know that, if we inadvertently precipitate a punctuation, life will go on in a new stable state. It is a near certainty that the new state will be less favorable for humans than the one we now enjoy."[33] So perhaps Godzilla may be perceived as the unstoppable natural force, capable of maintaining planetary balance—perhaps even a Gaian negative feedback control incarnate at the ready when a stable planetary-scale ecosystem tends to go awry?

In 2017 Legendary cast their version of King Kong (*Kong: Skull Island*) as another force, this time balancing Skull Island's odd ecology—so threatening to *Homo sapiens* if left uncontrolled. And yet Godzilla and Kong inevitably collided (for a second time) in 2021, recalling Carl Sagan's musings of half a century before. Anthropoid Kong is our mammalian cousin—closely allied to Man's evolutionary branch. So shouldn't the good guy prevail? According to young Sam Brody, though, who simply perceives Godzilla—Serizawa's primordial alpha predator—as a giant dinosaur, "The dinosaur always wins."[34]

CHAPTER TWO

Dinosaur Movies and Cryptozoological, Anachronistic Prehistoric Monsters of Film and Literature (1853 to 1963)

The Path to King Kong
and/versus Gojira

Introduction

Dinosaurs of the imagination—paleo-monsters from the Id! There are several tried-and-true methods for creating a scary prehistoric monster with which to thrill movie audiences. So let's consider a number of developmental firsts and early pivotal technological stages, inevitably wending toward 1963's blockbuster *King Kong vs. Godzilla*—yes, *the* battle of the 20th-century movie that really rocked my younger world.

Accept this chapter both as backstory and an overview of what is to come.

Dino-Monster Erector Sets

The first standardized approach would be simply to build, or erect one ... well, not exactly "simply." It's very hard to do. Yet today, and for many, dinosaurs are necessary, and necessity is the mother of invention.

Here we must first recognize Benjamin Waterhouse Hawkins

(1807–1894), whose herculean efforts gave the world its first dramatic glimpse of how several of these antediluvian, Frankensteinian horrors *may* have (but actually didn't, as knowledge later accumulated) appeared in life. Hawkins was a sculptor and the nonpareil dinosaur artist of his time. While prior to Hawkins's ascendancy there were several creators of paleo-iconography, Hawkins's prominence rose particularly as the paleontological sciences rapidly matured during the 1850s through the 1870s, and his restorations remained influential for 40 years.

Thanks in part to Hawkins, this most productive period of his career may even be considered the dawn of an early renaissance phase in scientific understanding of dinosaurs and other prehistoric animals.[1] Hawkins's life-sized sculptures of Mesozoic monsters, especially those constructed from concrete and iron bar for the Crystal Palace grounds at Sydenham, England—still standing today—are far from accurate based on current knowledge, but they were the first such made and for their time were highly influential, pop-culturally.

Over the past twelve decades numerous life-sized prehistoric animal representations have been sculpted (some animatronically designed) and scattered widely over the globe, confronting patrons of natural history museums, malls, roadside dinosaur theme parks, or even those occasional World's Fair prehistoric world displays. Life-sized dinosaur statues appeared in one of the earliest documentary films—Max Fleischer's *Evolution* (1923). *Evolution* not only inserted stop-motion animated dinosaur battle scenes filmed by Willis O'Brien and Herbert M. Dawley for *The Ghost of Slumber Mountain* (1919), but also featured footage of dinosaur sculptures made by Josef Pallenberg during the early 1900s for the Hamburg Zoo.

Today, life-sized dinosaur statues such as Pallenberg's are relatively common. But interestingly, Hawkins's most famous, earlier sculptures were of Mesozoic paleo-monster types, essentially foreshadowing giant movie dino-monsters projected on camera by Toho a century later in the years leading toward *King Kong vs. Godzilla*. Hawkins's renditions of pterodactyls (e.g., cinematic Rodan), an armored *Hylaeosaurus* (e.g., "Anguirus" and "Varan"), the *Iguanodon* and a theropod—*Megalosaurus* (e.g., related to *T. rex*, a carnivorous plus biped dinosaurian persona later influential toward and melded into the Godzilla suit design)—were completed for the Crystal Palace's grand opening in June 1854. Film historians know as well that throughout the history of sci-fi dino-monster making, life-sized, static three-dimensional representations of dinosaurs (e.g., skeletons and life-scale sculptural figures) have occasionally been relied upon as special-effect movie props.

Dino-Monsters Spawned from Canvas Paleoart

Eventually, other paleoartists would supersede Hawkins as ever more brilliant, go-to artists of prehistoric life—Charles R. Knight (1874–1953), Rudolph Zallinger (1919–1995) and Zdenek Burian (1905–1981) included among the most famous. Their paintings were vastly fortified with scientific knowledge previously unknown to Hawkins during the 1850s. Knight's foundational work in particular was often copied, and was supremely influential in the design of dinosaur puppets sculpted for 1925's *The Lost World* and *King Kong* (1933).

The countenance of Knight's 1897 painted restoration of *Agathaumas*, a real horned dinosaur showcased in 1925's *The Lost World*, grafted onto Anguirus's ankylosaurian body, is recognizable within its thorny-head, as well. Features of Zallinger's pot-bellied *Tyrannosaurus* for Yale's Peabody Museum *Age of Reptiles* mural—as represented in a 1953 issue of *Life* magazine, melded with one of Burian's *Iguanodon* restorations—were key ingredients captured within Godzilla's costume design. These three paleoartists had also completed *Stegosaurus* restorations, which Toho design artists must have seen—thus influencing that final icing on the cake addition—Godzilla's enormous, iconic spinal osteoderms. (Interestingly, in Ib Melchoir's and Ed Watson's unfilmed [circa] 1958 script to *The Volcano Monsters*, Godzilla and Anguirus were to be recast not as daikaiju, but as alleged dinosaurs—pitting a theropod tyrannosaur [i.e., Godzilla] versus an ankylosaur [i.e., Anguirus], battling to the death in San Francisco.)

Literary Dino-Monsters

Well, most dino-monsters couldn't read or write, but another way to captivate the masses with dino-monsters is through *writing about* them!

Beyond early sculptures and images of dinosaurs and other prehistoric animals were those amazing stories and literature dreamed up by writers, practitioners of science fiction and fantasy, involving Man's direct encounters with dino-monsters in misty, or forbidden primeval settings. A few writers, including scientists, had written formerly (although now generally obscure) popular works on paleontology, sometimes incorporating fascinating, speculative passages. However, it was Jules Verne (1828–1905) who wrote the first such celebrated genre tale, a novel invoking the menace of prehistoric life, *Journey to the Center of the Earth*, first published in 1864 in French, with a later 1867 French edition.[2]

Verne's *Journey* takes pride of place here for its Chapter 33, describing marine combat between two (much larger than in life, as known from fossils) 100-foot-long sea monsters fiendishly out of geological time—the *Plesiosaurus* versus its mortal enemy, *Ichthyosaurus*. Ultimately, the former succumbs to *Ichthyosaurus*, as Verne's intrepid explorers continue their journey through Earth's subterranean caverns on the trail of 16th-century alchemist, Arne Saknussemm. Thus, Verne anticipated the thrill and excitement imaginative audiences would have viewing gigantic reptilian dino-monsters battling other creatures from the recesses of prehistory (decades later on a movie screen).

Another strange tie-in may be made to an 1872 English translation of Verne's *Journey*. Evidently, an unknown author (possibly the translator) inserted his (or her) own chapter into the novel as Chapter 40, "The Ape Gigans." Verne scholars denounce this addition, although it's rather exciting and especially intriguing because it involves a large prehistoric gorilla (presaging Kong's entry sixty years later) that battles the equally monstrous shark-crocodile. Ape Gigans subdues its ancient nemesis, although the sequence turns out to be a letdown; yes, it's only a nightmare.[3]

In time this curious genre of literature, inaugurated by Verne, would encompass dino-monsters—often attacking civilized centers of the world—of every imaginable persuasion, including those from all manner of lost worlds, outer space, or those resurrected from fossil DNA via bioengineering. Imagination knows no bounds!

Cartoon-ish Dino-Monsters

In an early merging of paleoimagery with writing (melded as "imagetext"), magazine editors, book publishers and eventually movie producers realized that cartoons and comics could offer keen perspectives on dino-monsters, too, thus introducing the two-dimensional animated dino-monster (i.e., drawn on paper by artists—not filmed relying on 3-D articulated, miniature puppets and sculptures). Enter the cartoon-ish dino-monster!

It's difficult to pin down exactly when the first such single-panel comics and fantasy cartoons incorporating dino-monsters were seen in print, but several interesting examples of artwork were already appearing during the 1850s in publications like *Punch* and thereafter, often relying on Hawkins's dino-monster designs.[4] For our purposes, though, I nod to 1886, when a what-if type sketch of a gigantic, oversized *Iguanodon* bracing itself against a then-modern skyscraper

appeared in a book written by Camille Flammarion. More such startling images appeared, such as another 1886 drawing published by
Flammarion showing an early Stegosaurian variant dino-monster peering through a 6th-floor window in that author's marvelous presentation
of life-through-geological-time, a volume titled *Le monde avant la creation de l'homme*.[5] The fantastic paleoimagery theme was taken up once
more in an 1898 *New York Journal* showing a living *Brontosaurus* tanta-
lizingly gazing through an 11th-story window![6]

**Figure 6: An early restoration of the Stegosaurus-brontosaur chi-
mera appearing in Camille Flammarion's *Le monde avant la creation de
l'homme* (1886), anticipating the thrilling moments of giant monster films
that would come decades later.**

Of course, such compositional subject matter would become ever more popular in early 20th-century newsprint as well. Dino-monsters in speculative and science fictionally themed cartoons of every persuasion really began taking off then. And here we briefly turn to artist Winsor McCay (1867–1934), famous for his *Gertie the Dinosaur* (1912, 1914) silent cartoon, and an artist subject to Ulrich Merkl's lavish 2015 treatise, *Dinomania: The Lost Art of Winsor McCay, the Secret Origins of King Kong, and the Urge to Destroy New York*.[7] Merkl suggests that before his death in 1934, McCay intended to carry the ball much further, through a newspaper comic strip involving a larger-than-life *Brontosaurus* named Dino, which after being resurrected from a quarry wends its way through a metropolitan center, smashing things along the route. Was the idea behind Dino a precursor of sorts, or more intriguingly a possible influencer of giant, fierce dino-monster on the loose in tales and films that became so popular in following decades?

Alas, despite its promise, McCay's grand Dino adventure project never reached fruition (although Merkl has uncovered several draft 1934 illustrations and panels intended for the comic strip, published in his book). Prints of McCay's pairing of a giant monster on the rampage and animated cartoons (*Gertie on Tour* and *The Pet*, produced by 1921) were rarely circulated and "have sunk practically without a trace." However, Merkl opines that "The Pet," especially, which "tells an apocalyptic tale of a man who ... dreams that his wife's adorable new puppy-like pet grows into a beast of monstrous proportions, roams through skyscrapers King Kong–like, and eventually requires a platoon of airplanes to bomb and destroy it," preceded the climactic scene in *The Lost World*, and foreshadowed what was to come decades later in the movie industry. Merkl concludes: "Due to an injustice of history was not Winsor McCay's *The Pet* and *Gertie on Tour* in 1921, but *The Lost World* 1925, and then *King Kong* in 1933, that launched the giant beast or giant monster subgenre of science fiction, inspiring the atomic mutants of the 1950s, Japanese movie like Godzilla, and all the other big-screen creatures."[8]

Strange, peculiar dino-monsters increasingly became more commonplace denizens of cartoon animation. During the years spanning 1933's *King Kong* and 1954's *Gojira*, two stand out as exemplary. A prolonged segment in Disney's "Rite of Spring" (*Fantasia*, produced by Walt Disney and Ben Sharpsteen, 1940) offers glimpses of Earth's evolutionary life through time, climaxing with an epic dinosaur battle. Such raw emotional ferocity is conveyed through this *Tyrannosaurus* vs. *Stegosaurus* fight—the most intense dino-monster conflict audiences had witnessed since *King Kong*! Secondly, in a 1942 Paramount production directed by Max Fleischer, Superman fought the *Arctic*

Giant, a 100-foot-tall Tyrannosaur found frozen in Siberian ice. When it thaws in Metropolis's Museum of Natural History, the huge, finned pre-Godzillean dino-monster ravages the city, until Superman saves the day (and, of course, *Daily Planet* reporter Lois Lane). Looking further ahead, the year 1988 brought us *The Land Before Time* (directed by Don Bluth), so entertaining with juvenile audiences.

Eventually, besides superheroes, dinos in comic books became all the rage: V.T. Hamlin's *Alley Oop*, DC's *Tor* and Western Publishing Company's *Turok Son of Stone* earning pride of place. Donald F. Glut discusses these titles and many more cartoons and comics in his 1993 film *Dinosaur Movies*, as well as in Chapters 8 and 10 of his illustrious 1980 book *The Dinosaur Scrapbook*. As Glut notes: "Readers of Stone-Age-setting comic books *expect* to find dinosaurs; otherwise they find the stories mundane, in a medium in which spectacle and fantastic adventure are commonplace. ... Doubtless, as long as there is a comic book industry, the dinosaur and its primitive 'kin' will not be extinct in their four-color pages."[9] From there it's only a short conceptual leap to outright prehistoric monster madness, as in DC's *The War That Time Forgot* (with its daikaiju-like dinos and giant prehistoric apes),[10] Charlton's *Gorgo*, and Marvel's or Dark Horse's *Godzilla* dino-monster comic books.

Robotic, Mechanical Recreations

There were and still are every persuasion of large model, animatronic dino-monsters as well in the movies. Perhaps the earliest such creation was a life-sized *Ceratosaurus* appearing in 1913's *Brute Force* (directed by D.W. Griffith), yet a simplified structural example that opened its mouth and tilted forward down and up on a fulcrum. While far more sophisticated (electronic) animatronics—some quite expensive—have been used off and on since the late 1910s through the 1990s (e.g., *T. rexes* constructed and filmed for 1993's *Carnosaur*, directed by Adam Simon, and *Jurassic Park*, directed by Steven Spielberg), such robotic dino-monsters are seldom used as mechanical movie props; instead, other specialized methods are preferred by artists, technicians, or even actors.

Stop-motion Dino-puppets

Now let's survey evolution of the 3-D film technique as it pertains to the science fictional subgenre of onscreen prehistoric monsters, focusing on key origination stages—beginning at a suitable beginning with development of stop-motion animated dino-puppets.

By the early 1950s, movies relying on three-dimensional, articulated puppets permitting stop-motion animation, use of live reptiles sporting prehistoric-looking prosthetics such as horns and fins, and suit costumes to be worn by human actors masquerading as giant prehistoric dino-monsters became the most favored form of bringing prehistoria back alive for audiences to cower in fear. With this said, note that there were also approaches rarely taken that were conceptually intermediate between the 2-D cartoon-ish and 3-D puppet animation stages, using "jointed cut-out" dinosaur figures (presumably made of cardboard or wood) subjected to stop-motion animation, such as in a German c. 1927 production, *Ein Ruckblic in die Urwelt.*

Kong—it just doesn't get any better than this, many (oldsters like me) would proclaim! But movies like RKO's 1933 masterpiece just don't happen out of the blue. In fact, there was a two-decade lead-up to this pinnacle of filmdom. We should consider which movie(s) laid the groundwork thus grandfathering-in a perfected stop-motion dino-monster technique for Willis O'Brien (1886–1962), Marcel Delgado (1901–1976) and other craftsmen to utilize during the 1930s. So was the first 1933's *King Kong*, or an important film that came before? For proper perspective we briefly turn to paleoartist and dinosaur sci-fi film sculptor, Stephen A. Czerkas's (1951–2015) informative 2016 book *Major Herbert M. Dawley: An Artist's Life—Dinosaurs, Movies, Show-Biz, & Pierce-Arrow Automobiles.*[11]

Interestingly, both McCay and Dawley (1880–1970) were inspired, respectively, to their trick-photography craft of recreating "reel" dinosaurs appearing live on camera after (independently) viewing the American Museum of Natural History's *Brontosaurus* skeleton and display in New York. Charles Knight's miniature sculptural and painted restorations of this dinosaur accented its fossil bones in the museum exhibit.

Willis O'Brien was also smitten with the long-necked beast, but he soon sparred with Dawley. Turns out, their conflict and animosity toward one another had been misunderstood for decades, with Dawley cast as the villain. Yet as Czerkas elucidates, such characterization is unfair. Briefly, Czerkas's astute documentation revealed that both Dawley and O'Brien merit acknowledgment for developing the stop-motion animation trick-photography technique, as it applied to prehistoric themes (not O'Brien alone). In fact, when O'Brien collaborated on 1919's *The Ghost of Slumber Mountain* with Dawley—an artist whose techniques were more refined at the time—O'Brien was merely his apprentice. True—while prior to meeting Dawley, O'Brien had already produced a number of short stop-motion animation films on prehistoric themes dating back to 1915, although relying on relatively unsophisticated sculpted puppets. So, by 1919, the two collaborated on *The Ghost of Slumber Mountain*, aspiring for something far greater … beyond what they had individually achieved before.

Dawley's and O'Brien's mutual animosity, a falling out over proper use of lead credits as screened for *Ghost*, led to feuding and eventually legal bickering over rights to use the stop-motion technique, patented in July 1920 by Dawley as "Articulated Effigy." Their squabble escalated in the case of 1925's *The Lost World*—perhaps the earliest film which may be regarded as fortified with "refined" use of the lauded animation method involving dino-monsters (and apish suitmation, which we'll come to). Yes, as Czerkas documents, things really came to a boil with O'Brien's cinematic masterpiece, First National's *The Lost World* (1925) based on Arthur Conan Doyle's 1912 novel. Yes, and arguably this became *the* film where—in the case of cinematic dino-monsters—the stop-motion method really got launched.

Figure 7: One of Willis O'Brien's abandoned film projects was *The War Eagles*, of the late 1930s, which would have become another motion picture of the stop-motion animation variety. This drawing is suggestive of a scene in which tribal warriors of a lost civilization in the Arctic ride enormous eagles into battle against vicious allosaurs (illustration by Mike Fredericks [2019], editor of *Prehistoric Times* magazine, used with permission).

Judging from numerous video shorts posted on YouTube, stop-motion dinosaur/monster filming, relying on animated miniature articulated puppets, remained *the* most popular and revered technique through the Ray Harryhausen (1920–2013), Jim Danforth (b.1940), and Arthur Hayward (1889–1971), etc., era of practicing, into the 1980s, that is, until the advent of CGI—perhaps culminating in 1993's blockbuster *Jurassic Park*, yet extending through Legendary's giant monster series of the 2010s. Certainly, stop-motion animated sci-fi dino-monsters were also the most prolifically relied upon means for filming such creatures in a host of movies and other film shorts, from the 1910s through the 1980s.

Techniques refined for *The Lost World* then could be eventually relied upon for *King Kong* and 1953's *The Beast from 20,000 Fathoms* (directed by Eugene Lourie), both of which provided cinematic fuel for the most elaborate suitmation monster movie of the era—*Gojira* (1954). But before the unrivaled ascent of Japanese suitmation—the method Toho specialized in—dino-monsters, movie monster makers turned to another technique, that is, relying on live reptiles enlarged onscreen to dinosaurian proportions.

Unreal Dinosaur Masquerade on Reel

By the 1930s, before avian-dinosaur evolutionary connections were established, the general public had commonly thought of dinosaurs as former gigantic lizards and reptiles. So wasn't it a natural extrapolation to create living pseudo-dino-movie-monsters by adding fins, horns, and other prosthetics to live reptiles, thus rendering their appearances dinosaurian?

This trick had been tried before in 1913's *Brute Force*, directed by D.W. Griffith (1875–1948) with a "cosmetically-altered snake and alligator," and also in a 1936 *Flash Gordon* segment, but in 1940's *One Million B.C.* (directed by Hal Roach [1882–1982] and Hal Roach, Jr.) studio artists went all out in creating a prehistoric extravaganza, among many exciting scenes, for instance, incorporating a dramatic dinosaur battle for the ages, pitting a dwarf alligator dressed as a quasi–*Dimetrodon* versus a menacing gigantic tegu lizard, both scaled-up through special effects magic. An impressive stegosaurian-plated and single-horned Chuck-Walla lizard haunts an opening scene. Slow-motion filming with scary, monstrous roars dubbed in created quite an audience experience. Co-producer Hal Roach's and D.W. Griffith's "gagged-up lizards" proved sufficiently "convincing as prehistoric monsters," as dinosaur expert Don Glut, who wrote the definitive article on this film for the

August 1966 issue of *Modern Monsters*, suggests in his 2001 book *Jurassic Classics*.[12]

On stage, live reptiles could be non-cooperative or finicky, as directors and the animals' handlers noted. Some filmed scenes, such as the *Dimetrodon* Central Sea shoreline sequence in 1959's *Journey to the Center of the Earth* (directed by Henry Levin), seemed magnificent, while others, such as those involving Irwin Allen's (1916–1991) "Irwinosaurs," seen in the 1960 version of *The Lost World*, would be regarded (by adults) as, at best, questionable denizens in any prehistoric world. And so, after two decades of variable success using lizards and reptiles masquerading as dino-monsters, the practice faded. Glut was probably the last movie producer/director to utilize a tegu lizard "Neecha," and also two other reptiles, appearing in his 1996 film *Dinosaur Valley Girls*: yes, I met his pet Neecha a quarter of a century ago while it was dozing in an aquarium tank. (I was declined an autograph from this saurian star basking under a heat lamp in his aquarium, though.)

Apish Suitmation

On the massive heels of *King Kong*'s extraordinary success, gorilla suits (ape-mation) used in movies and theatrical serials became all the rage! It was also a time of considerable intrigue lingering over Charles Darwin's (1809–1882) theory of evolution via natural selection. Hence the evolutionary missing-link theme, probing man's possible ties to the most monstrous-looking of the great apes, proved extremely popular in many films and movie serials. Aghast, people feared: was the gorilla/anthropoid monster genetically instilled within us? The titanic *Tyrannosaurus* versus three-horned *Triceratops* confrontation as envisioned in Charles R. Knight's magnificent museum paintings conveys a sense of the prehistoric world's immortal savagery, but the idea of an ape versus dino-monster showdown intuitively seems more visceral.

In *One Million B.C.*, besides those live reptiles and other mammals filmed as huge, hairy and prehistoric-looking, Hal Roach also employed suit actor Paul Stader who wore a rubber *Tyrannosaurus* costume, thus presaging the preferred use of suitmation in sci-fi dino-monster films, a technique soon to prosper. For the record, prehistoric animal suits and costumes used in movie dino-monster making predated *One Million B.C.*, ranging back to at least 1927.

However, factoring filmic ape suits and dino-monster costumes *collectively*, surely the 1930s through the 1950s became a heyday of prehistoric gorilla-monster ape suitmation.

This was evidently leading to an inevitable showdown—apes versus dino-monsters. Conceptually, a quarter century after RKO's masterpiece, the battle rejoined with 1948's *Unknown Island* (directed by Jack Bernhard). This lost world–ish tale employed some of the worst dino-monster (i.e., *Ceratosaurus*) suits seen in a motion picture. And there was also a giant prehistoric sloth *Megatherium*, enacted by veteran actor Ray "Crash" Corrigan (1902–1976) who instead improvised this role while wearing a gorilla suit, lacking the Ice Age genus's characteristic powerful and massive tail. So instead of what had been intended in the script—two genuine prehistoric monsters founded on 1942 paintings by Charles R. Knight completed for *National Geographic Magazine*—their filmed battle foreshadowed the look and feel of yet another metaphorical ape versus dinosaur conflict, to be discussed shortly. A decade earlier, Corrigan had worn a gorilla suit fitted with a horn on its forehead in a 1936 *Flash Gordon* segment.

Arguably, however, the first genuine gorilla-monster suit intended to represent a prehistoric ape was worn by actor Bull Montana in

Figure 8: While this scene may seem like it came from a Kong vs. Godzilla flick, instead it was filmed for 1948's *Unknown Island*. The gorilla-suited actor at right grappling with the actor wearing the horned ceratosaur suit at left is intended to be a giant ground sloth, although lacking its hallmark powerful tail.

The Lost World. According to Michael Klossner, author of *Prehistoric Humans in Film and Television* (2006), "*The Lost World* was the most important silent special effects film and introduced audiences to sights they had never seen before."[13] Analogies have been made between *Lost World*'s Ape-man and RKO's Kong. For instance, both appear to be the last of their respective species—which are extinguished by man. Bull Montana's costume is well designed, appearing "primitive with a simian face, big teeth, a nasty grimace and a powerful hairy body." Furthermore, "the Ape-man survives among hordes of dinosaurs."[14] Before Kong's imagined conception, this Ape-man may be regarded as sort of a junior Kong. Another gorilla-monster link between Man and ape, played in a suit by Charles Gemora who was doubled in more dangerous scenes by Joe Bonomo, was featured in the 1932 film version (directed by Robert Florey) of Edgar Allan Poe's short story *The Murders in the Rue Morgue* (1841, 1845), starring Bela Lugosi.

Following *Lost World*'s and, especially, *King Kong*'s successes, the use of monster-ape suits became very popular for two decades in Hollywood, extending to Toho's use of a shabby Kong suit in their *King Kong vs. Godzilla*. Note that, here, while emphasizing those apish primeval monster and gorilla costumes and suits of filmdom, paleo-anthropologically speaking, a *prehistoric* caveman isn't necessarily also a *primitive* ape-man.

Glut whimsically refers to some of those Kong-ish suit imitators in his *Classic Movie Monsters*[15] as "second banana Kongs" of the mid–20th century, prior to 1962, illustrating many such examples in his directed and produced 1995 film *Hollywood Goes Ape*. Besides Corrigan's several performances, actor Charles Gemora also often played monster gorillas in many films after donning the requisite costumes. Gemora even appeared as a giant gorilla in 1949's *Africa Screams* (directed by Charles Barton), starring Bud Abbott and Lou Costello. Another giant ape, played by Paul Stockman, preceding Toho's blockbuster showdown appeared in *Konga* (directed by John Lemont, 1961), wearing a gorilla suit made by George Barrows. Steve Calvert also worked often as a gorilla suit film actor during this period.

But while those gorilla ape suits prospered mightily for several decades of the mid–20th century—especially prior to 1953—dino-monster suits really didn't take off until the 1950s. While gorilla-ape suits had a much earlier origin of use in America in films, the *dino-monster* suitmation industry or success story doesn't really begin until Japanese company Toho scored early successes with 1954's *Gojira*.

"C-law"-suit Brawl

Unlike the monsters' final battle ending in a tie, clearly suit design for *King Kong vs. Godzilla* had a distinct winner and loser. Godzilla's was my personal favorite, scoring victory for suit design: Kong was an abject loser!

Here's what Steve Ryfle had to say about that sad 1962 Kong costume, which to my then-nine-year-old eyes was at best *highly* disappointing. Ryfle stated of Kong's "flea-bitten" suit that "the sculptors had a hard time coming up with a Kong design that pleased (Eiji) Tsuburaya. Their first Kong model was fat, with coarse hair and long legs and arms, looking more cute and cuddly than Willis O'Brien's horrific beast. ... two Kong costumes were built ... one had a flat head, short, human-length arms that enabled the actor inside to scratch his head, throw his shoulders, and wrestle Godzilla at close range, and perpetually open white eyes. The second one had a triangular head, elongated ape arms for chest-pounding, and moving eyelids that flutter and close. ... His shabby fur looks unkempt like an old rug; his dopey face, with bushy eyebrows and crooked teeth, hardly instills fear, worst of all, his dumpy physique resembles an out-of-shape, middle-aged man, with sagging chest and pot belly." Ryfle also noted: "That sub-par King Kong suit has a lot to do with the film's negative reputation."[16] Ouch!

John LeMay adds in his informative *Kong Unmade: The Lost Films of Skull Island* (2019) that sculptor "Sadamasa Arikawa's ... final version (with an ugly un-heroic face and saggy build) was decided upon. Rumors circulate that this was intentional, as Kong represented America in the film, while contrasting reports say the (i.e., Kong) suits (two were constructed) appear as such due to budget constraints."[17]

Prehistoric monster Kong is generally considered the most perfected, beloved paragon of stop-motion animation achieved. Many Japanese monster movie aficionados would prefer to dismiss Kong's 1962 suits, as they looked far worse than any of the American gorilla-monster ape suits preceding in movie serials of the 1930s and 1940s (or any made in the decades thereafter in the United States). Godzilla's 1954 entry, however, along with that of his first, 1955 sparring partner Anguirus, introduced the most elaborate and quintessential form of giant dino-monster suitmation performed until then and for many years thereafter. (Meanwhile, *Creature from the Black Lagoon's* 1954 Devonian Gillman costume represented a paragon for human-sized prehistoric monsters.)

Life Sparks Dino-Monster Suitmation in Japan

Since the early 1950s, Japan has specialized in suitmation, particularly in regard to its famous dino-monster menagerie, as opposed to other methods outlined here. At the outset for 1954's *Gojira*, Eiji Tsuburaya (1901–1970) would have preferred using stop-motion animation, as in RKO's revered *King Kong*, but he realized this technique would have been far too "expensive and time-consuming than *Godzilla*'s tight budget and schedule would permit."[18] So, reluctantly he employed the less sophisticated suitmation process, which as noted by Ryfle, "proved a more effective method to portray the kind of destruction Godzilla would become famous for."[19]

Nevertheless, for decades it remained rather a mystery to young giant monster fans, such as those reading *Famous Monsters of Filmland* back in the day, exactly how life was infused into the deadly Godzilla dino-daikaiju monster. Fortunately, through recent years there has been considerable documentation of how the earliest (1954 through 1962) Godzilla suits were created.

Ryfle, for instance, outlined the suitmation process initiated at Toho, leading toward the first Godzilla monster costume—a product of designer Kazuyoshi Abe's, sculptor Teizo Toshimitsu's (1909–1982) and art director Akira Watanabe's (1908–1999) artistic insights. As previously mentioned, Godzilla's suit design was founded on a Frankensteinian melding of three dinosaurs known to science—*Tyrannosaurus*, *Stegosaurus*, and *Iguanodon*—life restorations of which had been painted on canvas by renowned paleoartists such as Rudolph Zallinger, Charles R. Knight, and Zdenek Burian (the latter whose recreations proved influential toward stop-motion animated prehistoric animals witnessed in a wonderful 1954 Czech film, *Journey to the Beginning of Time*, Americanized in 1966).

Perhaps the single most influential painting, however, was Zallinger's portrayal of bipedal, carnivorous *Tyrannosaurus* as depicted on a 110-foot-long-by-16-foot-tall magnificent mural named *The Age of Reptiles* displayed at Yale University's Peabody Museum! Toho artists would have seen a representation of this mural in the September 7, 1953, issue of *Life* magazine. Godzilla's fins (which later in the process were made to cast a radiant glow when the monster's fiery ray issued from its maw) were borrowed from *Stegosaurus*'s spinal plates, while Godzilla's long, powerfully destructive arms (in contrast to the tyrannosaur's) may have been appropriated from bipedal reconstructions of the *Iguanodon*. Ryfle concluded, "In taking artistic license with evolutionary history, the designers were clearly more interested in creating something fantastic rather than realistic or logical, a spirit that marked the beginning of kaiju history."[20]

Five years later, Ed Godziszewski elaborated further in "The Making of Godzilla,"[21] detailing how the suits for *Gojira*'s sea monster were conceived and constructed. After the monster's dinosaurian nature was settled, Toshimitsu sculpted three designs in clay; a prototype version known as the "alligator version" for its skin texture was selected for build-up. Then in June 1954 the first man-sized suit was built for actor Haruo Nakajima (1929–2017) to test. After being sealed within— the suit looking just terrific—unfortunately, Nakajima found he could barely force the monster's legs and arms to move! The suit was too stiff; upon curing, the latex had become rock hard. So, a second suit was made that could be operated properly by Nakajima ... and the rest is history. (We will delve considerably further into the making of the first Godzilla costume in Chapter Five.)

Yet another suit was made for Godzilla's next appearance in 1955's *Godzilla Raids Again*, or, as I originally saw this film under an alternate title, *Gigantis the Fire Monster* (1959), which also featured quadruped Anguirus. But while I immensely enjoyed Americanized *Gigantis*, the new (3rd) Godzilla suit and this pairing of movies are generally not favored by kaiju fans. Ryfle comments, "The new Godzilla design ... created by.... Toshimitsu is ... more slender ... particularly in the lower body and legs, making the monster appear less powerful than before. In addition, the costume's arms, claws and tail betray that it was constructed with a cloth base over which latex was applied ... tends to bunch up like a baggy shirt at times." Furthermore, its head "has large ears and thick eyebrow ridges that detract from Godzilla's menace." But in contrast, Anguirus's appearance was an "inspired creation." "Toho would create many more quadrupedal monsters in future films, but never again would the illusion of a four-legged creature be so effectively done."[22] Shortly, too, during this interval Toho released *Rodan, the Flying Monster*, featuring another remarkable feat of prehistoric monster suitmation.

A few years would pass, during which 1960's suitmation *Gorgo* (directed by Eugene Lourie) would thrill younger movie audiences, eventually leading to the Godzilla suit witnessed in Toho's Americanized *King Kong vs. Godzilla*. This suit, known as the "kin-goji" design, remains my favorite of all Godzilla's filmic manifestations. Perhaps what attracted me most to "kin-goji" was that I was a young dinosaur-loving kid, and this Godzilla appeared more like a weird sort of "real" monstrous dinosaur than any other of its filmic ilk. If only any real dinosaur known to paleontologists—and that was a word I knew then—had looked like this imaginary one!

Of kin-goji, Ryfle stated that this

> Godzilla suit is one of the best in the series—more bulky, ominous, and powerful than the streamlined suit deployed in *Godzilla Raids Again* (U.S. director and film editor Hugo Grimaldi) and recapturing the monolithic proportions of the first Godzilla, while still flexible enough to allow suit actor Haruo Nakajima to perform full-body rolls and other very physical stunts. ... When viewed from a profile. ... Godzilla looks absolutely menacing. ... This particular incarnation of Godzilla has remained a worldwide fan favorite ... and its design was the basis for a popular Godzilla model kit sold in the U.S. by Aurora for many years.[23]

Suitmation success!

And Beyond

We note that during the first approxmiately 60 years following Hawkins's early visions of how antediluvian fossil dinosaurians may have appeared, most of the standard innovations and tricks for resurrecting such denizens from primeval times into fantastic and science fictional settings had been invented and tried—sort of an early flowering Cambrian explosion of artistic efforts and technological breakthroughs. Essentially, there have been relatively few developments in manifesting dino-monsters for public meditations since the 1920s, other than the 1990s invention of computer graphic special effects (CGI) and a refinement of suitmation acting methods.

A horde of computer-animated dino-monsters of every persuasion invaded public consciousness when *Jurassic Park* (1993) stormed our theaters—the franchise later introducing a bioengineered kaiju called *Indominus rex* in 2015's follow-up *Jurassic World*. If *King Kong vs. Godzilla*—promoted so enthusiastically through studio and advertising hype—was intended to be the battle of the (last) century, then surely 2021's *Godzilla vs. Kong*, enriched with its many wondrous computerized enhancements, may be the new millennium's all-time tussle royale! Postponements of this long-awaited film left fans mystified as to plot details and outcome of the battle. As of late 2020 fans could only salivate and sate ourselves with this note posted on the relevant IMDb page:

> In a new world where man and monsters now coexist, (Project) Monarch must lead the way to a prosperous future alongside the titans, keeping humanity in check. However, rival factions that want to manipulate the Titans for war begin to rise under the guise of a nefarious conspiracy, threatening to wipe out all life on the planet. Meanwhile on Skull Island, strange seismic activity draws the attention of Godzilla and Kong alike.

But then, almost miraculously, given what was then happening in the world and to the movie industry in general, the film finally appeared! (More on this to follow in Chapter Nine.)

Have we exhausted all possible artistic and technological capabilities for bringing dino-monsters to the big screen? It may seem so today, but human imagination knows no bounds, as one may ascertain from this overview of over a century's worth of historical periods in prehistoric monster imagineering.

CHAPTER THREE

Merciless Gaia
in Geological Time

"Earth is ... sick with the high technological fever of
mankind. ... Perhaps the biosphere will be knocked
into a new ice age; maybe the Earth will overheat as
the greenhouse effect puts us past the point of human
habitation. The breaking of biospheric malaria—like
fever alternating with chills—could give way to a fro-
zen or jungle world without human beings."
—Dorion Sagan in *Biospheres: Metamorphosis*
of Planet Earth (1990), p. 166

"It is clear that the Gaia hypothesis is correct at some
level: Organisms do play an important role in the over-
all function of the Earth system. ... there is no rea-
son why we need go the way of the dinosaurs. We know
enough about the dangers of the Earth system to sur-
vive for longer than they did, if we can survive the dan-
ger that we pose to ourselves."
—Lee R. Kump, James F. Kasting,
Robert G. Crane in *The Earth System*,
3rd edition, pp. 29, 424 (2018)

"We have met the enemy, and he is us."
—"Pogo," cited in Isaac Asimov's and
Frederik Pohl's *Our Angry Earth*, 1991, p. 22

Not unlike fictional Godzilla, life—having existed for one-third the
age of the Universe—may seem immortal. With or without facilitation
from formidable daikaiju, geological history proves that Gaia can certainly take of herself! But, in theory, what is Gaia? Mother Earth practicing seriously tough love on all of her DNA-infused spawn, threatening

extinction rites on those which become too unruly? Can an entity that somehow insensibly controls or regulates conditions for life's continual existence—in the context of millions of microbial and multicellular species—be perceived as the essential, alpha and omega living organism? Without her evolving streams of children, is Gaia then non-living? How rare are extra–Gaian entities in the cosmos?

Those well-adjusted to the sciences will judge this summary too sketchy and brief for judicially addressing such an intriguing topic (which assuredly is the case). So, for those who thirst for more on the many faces of Gaia, please consult the provided notes and references (most of which are in popular vein).

As Isaac Asimov and Frederick Pohl commented in 1991, "We would all like to think that there was something—some benign and superior kind of *Something*—that would step in and save us from the things that are going wrong with our world."[1] For many throughout history, that "something" took the form of a supernatural deity or presence of one persuasion or another with certain doctrines applied; for others, controlling nature through application of scientific methods and resulting technology became another kind of ethereal cavalry that would rescue us from our plight when times got rough. Increasingly, many have turned to the Earth itself, christening a concept or living force today known to many as the ancient goddess Gaia. Because all Earthly life genetically evolved from a single common ancestor species billions of years ago, according to Darwinian Law, then wouldn't such recognition underscore that indeed Earth and its unified biosphere could be regarded holistically as an interrelated organism?

Our modern concept of Gaia was detected through scientific means—geology, biology and chemistry especially.

Gaia was an embryonic scientific concept decades before British atmospheric chemist James Lovelock (b. 1919), who, with Lynn Margulis (1938–2011), deduced its/her (i.e., Gaia's) existence and persistence through geological time. For example, during the late 18th century, Scottish geologist James Hutton (1726–1797) perceived Earth as a "living machine" or "organized body," as a planet which, cyclically, has maintained conditions for life, without "vestige of a beginning, no prospect of an end." His *Transactions of the Royal Society of Edinburgh* paper of 1785, "Theory of the Earth," and a 1795 treatise, *Theory of the Earth with Proofs and Illustrations*, set the course for "uniformitarian" geologists—gradualists who differed ideologically from catastrophist-leaning rock men. Analogizing Earth's complex hydrogeological cycle (akin to blood in a biological circulatory system), Hutton further claimed in 1785 that our planet "is a superorganism and that its proper study should be by physiology."[2]

Another "pre–Gaian thinker"[3] who in 1875 pondered Earth as a super-organism, referring to Earth's outermost concentric planetary layer as a "self-maintained biosphere," was Austrian geologist Edward Suess (1831–1914). Later, Russian mineralogist and geochemist Vladimir Ivanovitch Vernadsky (1863–1945) also contributed to Gaian thought and early modern conceptions of what is now known as Earth System Science, particularly through his 1926 book *Biosfera*. For Vernadsky, living matter is a "solar-powered mineral,"[4] the predominant geological force, inevitably increasing in potency through geological time! Others, as well, pondered the "fitness of the environment" for "physico-chemically" sustaining life, optimally through time.[5]

During the early 20th-century application of chemical thermodynamics turned to consideration of biological systems viewed through geological time on Spaceship Earth. Central to such developing ideologies was our view of a planet in which, based on the fossil record, (shifting) environmental conditions have (biogeochemically) remained suitable for (continually evolving) life over billions of years—while avoiding a true chemical equilibrium state that would have been inimical to life (as currently on Mars, for example). Life—having survived five mass extinction events of the geological past. Chemicals, which over the course of billions of years could have dissipated, but are necessary for living things, have remained more or less in balance (although steadily evolving in concentrations) through time, maintained by life itself, given the presence of nourishing liquid: water. The inseparability of environment *with* life (species), evolving as one through time—particularly regulated by microbials—was becoming recognized.

During the 1960s, consideration of the chemical composition of Earth's current atmosphere offered early, significant thermodynamic clues suggesting the existence of our Gaian entity. Gases that are ordinarily unstable together (as in a laboratory flask) should have reacted (equilibrated) long ago, reducing their chemical potential, instead of accumulating and maintaining what now appears as a well-balanced steady state of (unstable) atmospheric chemical disequilibrium. In fact, it is just such a strange, telltale admixture that would allow savvy, intelligent beings using proper instruments to detect life—Earth's metabolic system—from distant outer space. In particular, such a combustible, reactive (oxidizing) gas as oxygen should not have accumulated in the atmosphere (e.g., in the presence of abundant chemically-reduced carbon matter) based on mere chance. Oxygen we have become energetically adapted to should have rightfully been depleted eons ago, unless there was a *something* going on, interactively and continually manufacturing and replenishing it for life's self-preservation. Therefore, isn't Earth, including its rocks, minerals and inorganic substances, itself "alive" too? Gaia may be likened to a huge tree that is 99 percent dead wood and bark; yet the tree overall is living.

Lovelock stated, "Life and the environment are so closely coupled that evolution concerns Gaia, not the organisms or the environment taken separately."[6] Life does not violate the second law of thermodynamics. Indeed, life "has evolved with the Earth as a tightly coupled system so as to favor survival."[7] Once the spark of life is added, seemingly random and chaotic circumstances do equilibrate into a stable pattern—an ecosystem—over time.

Through time, then, the biosphere actively modifies and controls "its environment ... whose policy is always to turn existing conditions to its own advantage."[8] In fact, concentrations (partial pressures) of atmospheric gases appear to be in general balance with organisms that facilitate or mediate their production and consumption in a disparate variety of environments. Salinity of the oceans—from which life originated eons ago—has fostered living things ever since, maintaining dissolved ionic and elemental concentrations that will not disrupt cellular tissues. Lovelock has asserted that a reason for why the seas aren't saltier than about 3.4 percent is that "since life began, the salinity of the oceans has been under biological control."[9] Too much salt would kill off marine species and thus the seas—the physiological blood-line of the Earth system. And so Earth/Gaia has found a way of controlling excess salt concentrations in oceans through accumulation of coastal salt beds and formation of fossil salt formations.[10] Likewise, the pH of seawater has been amenable to life for millions of years.

If Earth exists in a habitable, just-right-for-living "Goldilocks Zone" it's partly because of Gaia's influence (predominantly via microorganisms intricately intertwined with the environment) on the ocean-atmosphere system, through geological time. However, through his technology and industry, mankind increasingly alters concentrations of several key (trace) gases, such as carbon dioxide and methane, further altering the equilibrium, causing hardships for species living in sensitive ecosystems across the globe. Since humans evolved naturally, we and our *anthropogenic* wastes and pollutants must also be critical components of the natural Earth system. And yet while Gaia *knows* how to persevere, it has no specific need of us—who evolved randomly, chaotically, according to Darwinian Law.

Gaia has no particular need of human life, or our pollution: life will go on regardless, as it has for eons, assimilating pollution into future resource materials. Assuredly, our myriad waste products scattered globally into every nook and cranny will be dealt with through longer-term recycling. True—the Gaia concept may *seem* comforting, but as Lovelock asserted, "People sometimes have the attitude that 'Gaia will look after us.' But that's wrong. ... Gaia will look after *herself*. And the best way for her to do that might well be to get rid of us."[11] (Here isn't Lovelock's concept of Gaia sounding rather like Medea?)

Furthermore, given our propensity for "cancerously" rapid population growth (i.e., the "population bomb"[12]), "No matter how adept an organism is at propagating itself, if it does not become involved in the life cycles of others, it will fail."[13] Or, as Lovelock surmised, "There can be no ... set of rules, for living within Gaia. For each of our different actions there are only consequences."[14] This is perhaps another way of stating that increased interdependency of biological diversity within ecosystems increases stability. Monoculture systems tend to fail.

Rachel Carson in her *Silent Spring* taught us about those dinosaur descendants, the birds and pesticides, but what's horrifically happening with the bees is only dawning. Bee populations are at risk and in decline, and if they vanished, without their pollination capabilities, according to Albert Einstein (apocryphally), "Man would only have four years of life left."[15] Their ranks have been thinned due to pesticide use (to ward off nuisance bugs), and now, thanks to climate change, murder hornets from the tropics are assaulting North American beehives. Are humans, in turn, simply nuisance mammals?

In 1988, Lovelock extended scope of Gaia's regulatory influences across geological time in his book *The Ages of Gaia: A Biography of Our Living Earth*. Here we learn of Gaia's origin.

> At some time early in the Earth's history before life existed, the solid Earth, the atmosphere, and oceans were still evolving by the laws of physics and chemistry alone. It was careering, downhill, to the lifeless steady state of a planet almost at equilibrium. Briefly, in its headlong flight through the ranges of chemical and physical states, it entered a stage favorable for life. At some special time in that stage, the newly formed living cells grew until their presence so affected the Earth's environment as to halt the headlong dive towards equilibrium. At that instant the living things, the rocks, the air, and the oceans merged to form the new entity, Gaia. Just as when the sperm merges with the egg, new life was conceived.[16]

However, Lovelock suspects that the origin of bacterial life on Earth preceded Gaia's emergence.[17] It took some indefinite time to become as-one.

Ever since, Earth's temperature—regulated to great degree, autonomically, by microbial-generated components of the ocean-atmosphere—has never veered so far in any direction (too hot or cold), on a planetary scale, such that life was obviated. In other words, life became the predominant geological force, increasing in scope and magnitude through time. Oceans persisted past 1.5 billion years ago, during the Archean Era, to the present as an indirect consequence of oxygen-generating photosynthesis and subsequent carbon depositional rock burial.[18] Lovelock believes that "Earth was saved from drying out by the abundance of water, and by the presence of Gaia, who acts to conserve water."[19]

Henceforth, during the Proterozoic Era (2.5 to 0.57 billion years ago), seas and atmosphere became increasingly oxygenated, poisoning (e.g., yes, rather in Medean fashion) colonies of prior ruling anaerobic organisms that had been flourishing. Through time, a series of life-stabilizing, intricately interwoven negative feedback systems—regulated by microbes and increasingly, as they evolved, by multicellular organisms as well—working in synchrony with geological processes, then prevented conditions on our primitive planet from adversely running away (like on Mars or Venus), trending toward bitter extremes, hostile to life. That's why from the depth of our solar system Earth appears as a "pale blue dot," to paraphrase Carl Sagan.

If Gaia's feedback systems are so intricately interwoven, reflecting actions of trillions upon trillions of organisms (micro- and macroscopic), how can one determine, quantitatively, whether or to what extent Earth systems have remained sufficiently balanced for life's persistence through (seemingly) interminable swaths of geological time? Well, first, one must generate reams of reliable real-world data. Then, such data may be input into sophisticated computer models (given key, underlying assumptions), which have been devised over the past half century. Even so, it is only possible to reliably model certain regulatory drivers of Earth's atmospheric system, such as oxygen and carbon dioxide levels through the ages.

One such model is named GEOCARB, modeling the carbon cycle of the past 550 million years, developed by a number of scientists headed by Robert Berner; another, more refined model is GEOCARBSULF. A host of long-term natural processes such as chemical weathering of silicate rocks, burial and weathering of organic matter, volcanic degassing of greenhouse gases—carbon dioxide and methane—and accumulation of oxygen in the atmosphere via photosynthesis, and many other factors must be considered. And then results may be calculated. Interestingly, such efforts do not rely on, or presume existence of, a Gaian entity

Figure 9: Earth from space showing the hues identifying our world as a living planet (NASA photograph).

or highlight major, yet geologically brief, perturbations to the Earth system (such as the asteroid impact of 66 million years ago—initially, a non–Gaian and non–Medean episode, since it stemmed from an extraterrestrial cause). Model analysis of projected atmospheric oxygen and carbon dioxide levels during the past 550 million years is performed. Researchers display the results graphically. Further striking conclusions and hypotheses may then be made and tested.[20]

In his *Out of Thin Air* (2006), paleontologist Peter D. Ward intriguingly traced the evolution of dinosaurs and birds in light of curves derived from the GEOCARBSULF model, indicating how atmospheric oxygen levels have varied through time. Several epochs of relatively low oxygen (e.g., 11 percent to 15 percent range)—or a period in which a sudden decline of 10 percent oxygen was recorded such as during the Permian-Triassic event—correspond to major mass extinction events of the past half-billion years. In particular, though, while fluctuating carbon dioxide held little correlation to "animal diversification rates,"[21] "change of oxygen levels from ... less than ... 15% ... stimulated new speciation rates."[22] In other words, necessitated by low atmospheric oxygen, lucky animals adapted in order to survive, and this was exactly the Triassic explosion environment during which earliest dinosaurs evolved (250 to 200 million years ago), resulting in their superior respiratory systems—later inherited by their evolutionary descendants, birds.[23] A subsequent period of low oxygen winnowed species even further, further favoring those with derived, or advanced, circulatory systems.

As Gaian theorists succinctly explain, a reason for why Earth's average overall temperature has stabilized through eons of geological time is that "Gaia is at work." In early planetary youth, "Earth's atmosphere contained more carbon dioxide that it does now ... plants came along to reduce the proportion of carbon dioxide in the air. As the Sun warmed up, the carbon dioxide, with its heat-retaining qualities, diminished—in exact step, over the millennia. Gaia worked through the plants ... to keep the world at the optimum temperature for life."[24]

In response to the Gaian self-sustaining, life-nurturing ideal, Ward countered with his lesser-known, contrasting Medea concept. Ward contends that the history of life does not support the Gaia hypothesis, thus falsifying it, while noting that this finding doesn't necessarily prove the correctness of Medean thought. However, "the evidence at hand certainly points to it being a better descriptor of how life works than the Gaia hypothesis."[25] Earth has certainly metamorphosed in time, and most remarkable are catastrophic episodes of the geological past—most sharply evident in the fossil record as faunal and floral extinctions and upheavals. Ward identified several mass poisoning/suicidal events of the past 3.5 billion years, when Gaia (or rather, metaphorical Medea) went rogue. If planetary Gaia seeks to maintain conditions for life, then what could possibly be beneficial about such calamitous periods in the history of life, which trend toward life's total extinction?

Goddess Medea certainly has rocked life backward on its heels several times, and for prolonged ages of time. Earth (and her living self with evolving cellular brood) has survived global methane and subsequent oxygen atmospheric poisonings (3.5 billion, and 2.45 billion years ago, respectively), and two incredibly long-lasting deep-freeze "Snowball Earths" (lasting from 2.4 to 2.1 billion years ago, and then again from 790 to 630 billion years ago). But somehow life endured, persevered ... found a way![26]

During the Phanerozoic, from the early Paleozoic Era onward through time toward the recent, widely swinging atmospheric conditions episodically arose when life needed to hang on ... for dear life. One of the most severe Medean episodes took place about 252 million years ago during an awful time known as the "Great Dying," or the Permian-Triassic transitional mass-extinction event. Subsequently, the fossil and rock record of the subsequent Mesozoic and early Cenozoic Eras reveals that *several* enhanced greenhouse conditions rapidly warmed oceans to such an extent that poisonous hydrogen sulfide gas bubbled to the surface (emitted by microorganisms normally found in deep ocean waters), causing widespread mortality on land. Oceans too became anoxic, resulting in further marine extinctions.[27]

By the 2000s Gaian science and philosophy yielded to a revolutionary paradigm. Stemming from such considerations, discrete scientific fields of study formerly referred to as Ecology and Geology, are now combined, or melded with other disciplines (e.g., Oceanography, Chemistry, Meteorology) into a more fully encompassing construct known as Earth System Science, reflecting in detail many of Lovelock's initiatives, quantitative analyses and keen insights, while not necessarily relying fully upon his theory as foundation.

Earth System Science now incorporates knowledge of fundamental geochemical processes on our mobile planet, such as the regulatory long-term influence of the carbonate-silicate rock cycle through time on atmospheric carbon dioxide content (including both terrestrial and marine organic cycles, as well as recycling of inorganic and organic carbon, inextricably linked within a *global* carbon cycle), studies of the recycling of other key elements such as nitrogen, sulfur, precious phosphorous, and ultimately consideration of climate change drivers through time—including man's recent industrial exertions and exhalations as a geologic force.

Virtually unknown half a century ago, today Gaia conjures philosophical, metaphorical, teleological and pseudo-scientific cogitations—she's a hard mother to pin down! During the 1980s, evolutionary biologist and science popularizer Stephen J. Gould (1941–2002) dismissed Gaia as mere metaphor. And as Professor of Philosophy Michael Ruse noted, "professional scientists hated Gaia ... also the general public loved it."[28]

Lovelock stated in 1991: "In many ways Gaia, like an invention is difficult to describe. The nearest I can reach is to say that Gaia is an evolving system, a system made up from all living things and their surface environment, the oceans, atmosphere, and crustal rocks, the two parts tightly coupled and indivisible. It is an 'emergent domain'—a system that has emerged from the reciprocal evolution of organisms and their environment over the eons of life on Earth."[29]

During the 1980s, Gaia theory came under fire, for if it/she is a true, holistic biological entity, then how would Gaia reproduce and how would ours and other competing Gaias evolve via natural selection—survival of the fittest? For in 1983 Richard Dawkins exposed what he perceived as the "fatal flaw" in Lovelock's theory, stating that in order to "strictly apply" the analogy, "there would have to have been a set of rival Gaias, presumably on different planets. Biospheres which did not develop efficient homeostatic regulation of their planetary atmospheres tended to go extinct. ... In addition, we would have to postulate some kind of reproduction, whereby successful planets spawned copies of their life forms on new planets."[30] Dawkins's criticism prompted Dorion Sagan, in rather a science fictional vein, to speculate that our Gaia is poised on a precipice (although perhaps not for millions of years) of reproducing intact, self-sustaining biospheres, casting its life into the cosmos, perhaps even seeding other worlds.[31] Panspermia perhaps? Are we (or rather, Earth-life in general) then not unlike Loren Eiseley's slime or virus, on the verge of seeding the stars?[32]

In order to perceive how actions characteristic to all biota might be possible from dim human perspective, the idea of a super-organismic goddess Gaia, beyond our full comprehension, is often relegated to the metaphorical, philosophical, or even pagan, teleological (New Age spiritualism, Green pro-nature environmental movement) realm, straying beyond what is scientifically falsifiable, especially from a sober, scientific perspective—unless such considerations can be validly incorporated within the Earth System. However, any means or mindsets that support diminishing wanton pollution, while reducing Man's carbon footprint and preventing the extinction of species in droves represents a positive, whatever one's cultural persuasion.

Charles Darwin became our first global ecologist, recognizing the significance of organisms with environment toward adaptive ecological survival. Later, Rachel Carson revealed how mankind in the industrial era was torturing Earth's ecological niches in her widely read, illuminating book *Silent Spring* (1962), emphasizing the questionable use of toxic chemicals in a massive, fruitless war against pestilent and nuisance insects. To Carson, the "parallel between and chemicals and radiation is exact and inescapable."[33] Carson's plea was underscored and echoed in Laurie Garrett's 1994 book *The Coming Plague: Newly Emerging Diseases in a World Out of Balance.*[34] Is now Gaia rebelling ... against *us*?

As stated by evolutionary paleontologist George Gaylord Simpson, "The major features of cellular organization, including, for instance, mitosis, must be much older than 500 million years—more than nearly 1,000 million. ... In this sense the world of life, while surely fragile and complex, is incredibly durable through time—more durable than mountains. This durability is wholly dependent on the almost incredible accuracy with which the inherited information is copied from generation to generation."[35] (Since 1957, when Simpson wrote those words, the origin of Earth's life has been extended back to well over 3 billion years.)

So, Gaia (or Medea) has existed for some 3.5 billion years, unconsciously transforming and evolving in concert with millions of species thriving or undergoing reduction in their ranks at certain, harrowing episodes in the history of life. But it has only been within the past ("interglacial") sliver of time, since the recent Ice Age, that one (or several) species evolved, graced with a distinctive capability to comprehend—to learn and to know better.

And through our transcendence, unconscious Gaia became *aware*.

CHAPTER FOUR

Kong-frontations
Introducing King Kong

Ever since the mid–19th-century discovery of the (lowland) African gorilla, this ordinarily gentle yet powerful beast-animal fueled fervid lines of thought, particularly with respect to human evolutionary origins, and of outright conflict within the animal world. For a time, the gorilla was considered "the most fantastic beast in the annals of natural history."[1] Richard Owen (1804–1892), who named the "Dinosauria" in 1842, referred to the gorilla as "the most strange and extraordinary animal of the brute creation."[2]

How did this reserved Paleolithic survival—the gorilla—since transform into a fearsome monster as witnessed by a litany of fantastic literary and filmic representations in the human imagination? And how have such representations evolved through time … into Kong, a "reminder of the bestial side of ourselves"[3] and beyond?

Hunted by man in modernity, large, powerful apes soon began haunting works of paleo-fiction, such as an English "bowdlerized" 1872 translation of Jules Verne's *Journey to the Center of the Earth*. That creature, named the "Ape-Gigans," inserted within Verne's classic by an author unknown today, which we'll soon return to, represents science fiction's first take on a battling Kong-like creature. Eventually, from 1933 to the present day, Kong and other quasi-Kongs of both fantasy literature and film extended the myth of a primeval monster allied to our ancestry that defeated giant reptiles and dino-monsters in mortal combat, further illustrating Carl Sagan's metaphorical "nighttime stirring of the dream dragons." Or this is a conflict that University of Chicago scholar W.J.T. Mitchell characterized as evoking "the ancient antagonism between the cold and the warm, reptile and mammal, dry bones and living creatures."[4] Donald Glut understood the vulgar side of this theme we know and love so well when he titled one of his DVDs

Dinosaurs vs. Apes (Frontline Entertainment, 2007). Let's survey some of these most memorable fights between Godzilla and Kong, viewed in the face of eco-balancing Gaia.

Because killing was then considered an acceptable means for obtaining specimens for zoological study—thus furthering knowledge of natural history—the gorilla's (e.g., morphologically, fictional Kong's closest real animal inspiration) earliest battles were waged (in an inverted *Planet of the Apes* scenario) with human hunters, as recorded by explorer Paul Du Chaillu (1831–1903) in his fascinating 1861 account *Explorations and Adventures in Equatorial Africa*. The native Mpongwe, facilitating Du Chaillu's travels, who named gorillas "Ngina" (also "njena" or "nguyla"), even believed that there was a "kind of gorilla—known to the initiated by mysterious signs, but chiefly by being of extraordinary size—which is the residence of departed negroes."[5] It was a traditional belief that a "possessed" male gorilla of the myth sometimes even forced tribal women to "submit to his desires."[6]

Du Chaillu documented his expedition's first gorilla "kill," stating, "Nearly six feet high ... with immense body, huge chest, and great muscular arms, with fiercely-glaring large deep gray eyes, and a hellish expression of face, which seemed to me like some nightmare vision: thus stood before us this king of the African forests. ... The roar of the gorilla is the most singular and awful noise heard in these African woods."[7] Gunfire smote this beast and several other specimens along Du Chaillu's route through Africa. Du Chaillu did regard the adults sometimes as "monster gorillas."[8] Du Chaillu was further aware of biological implications posed by the gorilla, noting paleontologist Richard Owen's view based on comparative anatomy, that the gorilla, whose skeletal remains had been known to science since 1847, was closer or "nearer akin to man" than the chimpanzee. Where was the definitive demarcation between man and ape; could one become the other, people both wondered and feared? Indeed, Du Chaillu wrote of his gorilla hunting, "It was as though I had killed some monstrous creation, which yet had something of humanity in it."[9] Later, Du Chaillu took a stuffed gorilla specimen to Broadway, creating a sensation, thus attracting the attention of P.T. Barnum! Many wondered whether the gorilla was the link between man and monkey. (The gorilla was, however, certainly far less related to the cold-blooded sauria.)

The possibility that the unseemly gorilla was somehow "closer to man" (as further illustrated in an intriguing 1863 diagram showing comparative skeletal anatomies by Benjamin Waterhouse Hawkins prepared as the frontispiece for Thomas Henry Huxley's [1825–1895] book *Man's Place in Nature*[10]) proved disturbing to Victorian audiences, who

Figure 10: "The death of my hunter." An image suggesting savagery of the gorilla appearing in Paul Du Chaillu's *Explorations and Adventures in Equatorial Africa* (1861).

ultimately feared the primeval animal beast within. As evolutionary debates further swept up the populace through the early 20th century, including one famous evolutionary debate held at Oxford University, England, in 1860 and then another known as the 1925 Scopes Trial held in Tennessee, writers, artists and showmen alike thrilled laymen with their fantastic views of prehistoric ape-like ancestors, sometimes anachronistically interacting with modern humans.[11]

For instance, Jules Verne penned passages about apparently living fossils of prehistoric men found in the sublime, subterranean Earth for the 1867 edition of his *Journey*.[12] Half a century later, Edgar Rice Burroughs (1875–1950) published *Tarzan of the Apes* in 1912 (the same year that Arthur Conan Doyle's *The Lost World*, replete with its fictional proto-men, appeared in *The Strand*[13]). In Burroughs's timeless tale, a human child is raised in the African jungle by a tribe of powerful, anthropoid Great Apes—further bridging the gap between apes and men. Visual representations of ape-like, savage proto-men, evolutionarily linked both to the gorilla and ourselves in modernity, left startling, lasting impressions. Gorillas were among the most popular displays in lecture halls and, later, zoos and other circus exhibitions—although there in such settings, not quite living up to their supposedly ferocious, monstrous reputations.

Increasingly during the late 19th and early 20th centuries, iconography and imagetext focused on evolutionary aspects of the gorilla

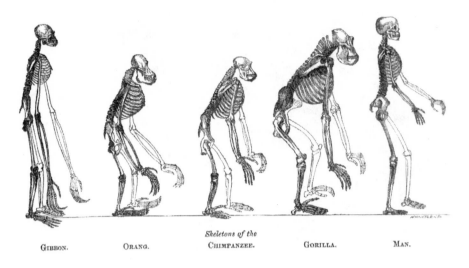

Skeletons of the

GIBBON.   ORANG.   CHIMPANZEE.   GORILLA.   MAN.

Figure 11: Benjamin Waterhouse Hawkins's diagram suggestive of an evolutionary trend of the primates leading toward exalted, upright man. Hawkins was not an evolutionist of his day, but this image appeared as a frontispiece to "Darwin's Bulldog" in Thomas Henry Huxley's 1863 book *Evidence as to Man's Place in Nature*.

and plausible kin. Popular writers besides Verne, Burroughs and Doyle imagined what must have seemed to their reading audiences, quite nearly the *unimaginable*—prehistoric, only partially evolved ape-like men who, in time, would transmute into modern humans. This common theme was explored by Jack London (1876–1916) in *Before Adam* (1907), Pierre Boitard (1789–1859) in *Paris Before Man* (1861), Charles G.D. Roberts (1860–1943) in *In the Morning of Time* (1919), and visual artists as well such as Charles R. Knight.[14] Stop-motion animation film legend Willis O'Brien's 1917 film *The Dinosaur and the Missing Link: A Prehistoric Tragedy* (Edison, Conquest Pictures) portrayed quite possibly the first saurian (i.e., a brontosaur) versus apish-gorilla battle in filmic history! As stated by Paul A. Woods, this short movie "offered a sneak preview of the battle far ahead of the fierce tyrannosaurus and Kong."[15]

So now on to those Kong-frontations! Here we find that while Godzilla has always merely been the up-and-coming contender, Kong, metaphorically, was long-undisputed champ! And in order to win in this arena, you must beat the champ. While Godzilla (or "Gigantis") began its path to the storied title bout by scoring a narrow victory over the Anguirus dino-monster in 1955,[16] Kong was already the long-undisputed king of Skull Island.

A century and a half ago, that unknown author who in 1872 penned a new Chapter Forty, inserted into an English edition of Verne's novel *A Journey to the Center of the Earth*, arguably created our first Kong, albeit named the fourteen-foot-high, coarse-haired Ape Gigans, which battles the "Shark Crocodile" in a volcanic shaft.[17] This is the first confrontation between giant ape and reptilian monster that I am aware of in science fiction. Literary purists denounce this desecration of Verne's original work, although in its own right the added passage, well, to me, reads delightfully. Ape Gigans is billed as the "antediluvian gorilla," the "progenitor of the hideous monster in Africa." It groans with the sound of "fifty bears in a fight," while its teeth are "like a mammoth saw." Its challenger, the Shark Crocodile—known to the "early writers on geology," possibly a reference to centuries-old writings on fossil shark teeth then referred to as "tongue-stones"—has a whale-sized body resembling the crocodile and a shark-like mouth. Ape Gigans defeats its reptilian nemesis, and then frighteningly approaches the story's protagonist, when suddenly "Harry" (i.e., "Axel" in prior French editions) awakens from his "nightmare sleep."[18] That's right, the Ape Gigans incident was only a dream!

Over six decades would pass before another giant gorilla-ape would battle ancient dominant reptiles—dinosaurs—for supremacy, reenacting the primordial conflict between mammals (i.e., metaphorically "us") versus our scaly, ancestral Earth-dominating nemeses, setting a trend for much of what would come thereafter. Chronologically, Kong's first official battles took place in print ... in 1932 via Delos W. Lovelace's (1894–1967) novelization from the *King Kong* movie script. Kong's creation gestated in Merian C. Cooper's (1893–1973) mind as a melding of W. Douglas Burden's (1898–1978) 1920s expeditions to the island of Komodo where giant (saurian) Komodo Dragons thrive, with Cooper's visions of a gigantic terror gorilla on the rampage—dominating its primeval foes, yet utterly lost in modernity.

Lovelace's engaging novel was commissioned to promote the then forthcoming 1933 RKO film, directed by Merian C. Cooper and Ernest B. Schoedsack (1893–1979). The set of anachronistically prehistoric animals Kong battles against in the book differs somewhat from those seen in the movie, because last-minute changes were in flux and edits underway as the film's release date drew near.

The first dinosaurs Kong confronts in the novel are a pack of feisty three-horned *Triceratopses*, a scripted scene chopped from full filming.[19] (To date the only published images I've seen of this incredible confrontation are Richard Powers's sketchy illustrations in a 1976 [abridged] edition of Lovelace's novel *The Illustrated King Kong*.[20])

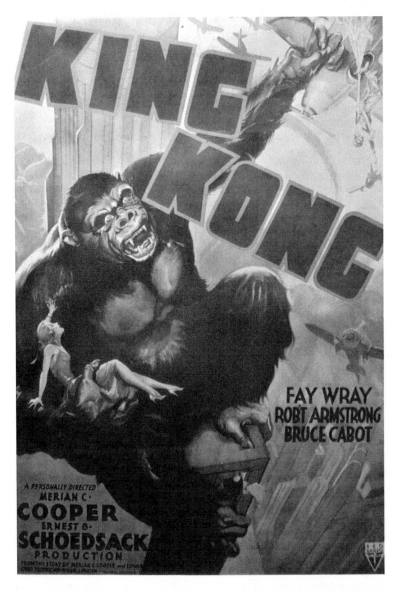

Figure 12: Perhaps the most famous movie poster for RKO's 1933 *King Kong* **(author's collection).**

Kong, still clutching Ann Darrow, is pursued by *Triceratopses* across an asphalt field (here, think La Brea Tar Pits). While some in the herd become mired, two continue to threaten Kong. So, the giant ape begins hurling large slabs of hardened asphalt toward the dinosaurs. "Kong cast his huge projectiles. One, striking fairly, broke off a horn. The

triceratops staggered, obviously hurt, and Kong redoubled his attack. ...
Kong roared in triumph and beat his breast."[21] Meanwhile the second
pursuing three-horn brute slowly retreats toward the band of watching explorers, as they drift in the direction of a ravine, over which a
fallen log bridge offered possibility of refuge. A crewman is gored by
this dino-monster, and in Lovelace's novel it is this particular *Triceratops*, now perceived as "hunting" them, that offers no way of survival
for the men other than to cross over the log, supposedly to safety. Too
bad this dramatic sequence was never filmed for the 1933 classic *Kong*!

The next creature Kong is menaced by, on the other side of the
ravine, is a *Tyrannosaurus*—arguably then and now science's most
famous dinosaur (although not named as such therein). However,
Lovelace's Rex is an antiquated, late–19th-century "hopping" version
of bipedal dinosaurian.[22] On film, this brutal brawl translated into
Kong's then most famous fight, yet one which in many instances was
read by audiences prior to what would shortly be projected onto the
silver screen theatrically via stop-motion animation. Kong defeats the

Figure 13: Kong showing his great strength by stuffing a tree into Tyrannosaurus's maw, a scene from 1933's *King Kong*, a scene repeated in several of
Kong's movie reappearances.

dino-monster, but we also read, "Any critical observer would have realized that Kong had met enemies of the meat-eater breed before and had worked out a technique of battle which served well when he was not too enraged to use it."[23] As apparent through his victory over the tyrannosaur, clearly Kong is undisputed lord of his domain!

Then, in the novel, Kong encounters two additional dinosaurian survivals from prehistory. The first is rather non-descript as to its true classification, yet we recognize it not as a snake, but as an aquatic elasmosaur.[24] Lovelace inserts this illuminating remark, forecasting Carl Sagan's, while Kong is entwined within its scaly coils. Kong "shivered, so full he was of the loathing his species has had of reptilian things since the dawn of time."[25] While atop his "lost world of Skull Mountain Island," a great pterodactyl swoops down to seize Ann, which of course is quickly dispatched. So many decades later, I still recall a caption penned by Forrest J Ackerman that was published in a once-owned 1960s vintage issue of *Famous Monsters of Filmland*, printed above a famous photo of Kong battling the huge *Pteranodon* on a cliff. Forrest J. Ackerman wrote, channeling the USA National Anthem, "Oh say can you see. By the Dawn's early light. That this Ptero-dac-tyl. Has flown its last flight."[26]

The 1933 film, released on March 2 in New York, thrilled audiences with several of these incredible stop-motion animated confrontations. The Kong versus *T. rex* fight is the individual movie scene that fused itself indelibly on my then-six-year-old brain when I first saw it (televised, circa 1959/60)—perhaps *the* dino-monster battle that all must be compared to. The (three-fingered) tyrannosaur puppet was sculpted by Marcel Delgado (1901–1976)—based upon magnificently painted museum restorations by Charles R. Knight. As Neil Pettigrew describes, the fight is "brilliantly sustained: both animals snarling, the dinosaur's tail swishing, the ape's fists pounding, both animals biting each other."[27] Steve Archer, one of Willis O'Brien's biographers had this to say about this scene in 1993: "O'Brien's experiences as a prize-fighter and his sharp observation of the way animals behave brought to the screen a fight whose actions have been often copied but never surpassed."[28]

In the end, our approximately 50-foot-tall Kong wins, after leaping onto the dinosaur's back and prying its toothy jaws apart, snapping the bones bloodily. Thumping his chest as if this is just another day in the red in tooth and claw jungle that is Skull Island, Kong seizes Ann and marches on. (Don Glut recreated the essence of this giant monster battle scene for his 1962 stop-motion animated amateur film *Tor, King of Beasts*.[29])

In Kong's cavernous, volcanic lair, a snaky, swimming elasmosaur slithers out of a pool, wrapping itself around the beast-god. This puppet doesn't resemble the actual genus now known to science, as it is far too snake-like. This isn't a prolonged battle like the previous one. Next, in a scene so complicated to animate—that according to Willis O'Brien took seven weeks just to film, working ten hours per day—supported by wires, a huge, toothed, winged *Pteranodon* (i.e., although the genus lacked teeth) descends toward the mountain peak, seizing Ann, intent on carrying her off maybe toward its nest. Pterodactyls were fragile-boned flyers, and so it doesn't take too much effort for Kong to tear into the leathery-winged monster.

There were several contemporary stories published for younger readers founded on either Lovelace's novelization or the film itself, involving giant apes. One advertisement for such a tale showing a large dryptosaur fighting a giant gorilla in London was published in an October 1933 issue of *Boys' Magazine*. (See Figure 1.) And primeval "giant gorilla" recreations inspired by *King Kong* sometimes made their way into thrilling animatronic prehistoric life exhibits, such as Messmore and Damon's "The World a Million Years Ago" exhibit at the Chicago World's Fair in 1933/34.[30]

Later in 1933, RKO came back with a rushed, 70-minute sequel titled *The Son of Kong* (directed by Ernest B. Schoedsack). The less mature, more diminutive (compared to the expired Kong), 12-foot-tall, white-furred gorilla-ape smashes a saurian—a long-necked dragonesque creature, otherwise unknown to paleontology. While entertaining and fortified with impressive stop-motion animated sequences, this movie is clearly not in the same league as the parent film. Also in 1933, *King Klunk*, a nine-minute cartoon by Walter Lantz (A Universal Cartoon), appeared. This obvious spoof of *King Kong* featured a battle between a hungry Rex-like saurian and the titular ape, Klunk. A boxing match ensues between them, in which the dinosaur delivers a buzz-sawing from its dorsal scales. After an around-the-world punch, Klunk literally flattens the Rex. (Donald Glut addresses a pop-cultural plethora of Kong-related imagery, recordings and publications—books, pulp magazines and comic book Kong-ish encounters—several of which involve battling tyrannosaurs in "His Majesty, *King Kong*—X: King Kong Lives."[31])

An intriguing fossil discovery was announced in 1935—that of the extinct *Gigantopithecus* (literally, "gigantic ape"), which is now known as a nine-foot-tall ourang-outang-like creature, non-ancestral to humans, formerly inhabiting South Asian forests from about one to ten million years ago! Some paleontologists suggest that early man (e.g., *Homo erectus*) may have exterminated *Gigantopithecus*. Meanwhile,

Kong—the king himself—retired for 30 years, although that didn't prevent other pretenders to the throne from appearing.

John LeMay and Ken Hollings independently refer to a Japanese Kong movie released during the late winter and spring of 1938 titled *Edo Ni Arawareta Kingu Kongu* (translated in English as "King Kong Appears in Edo").[32] This was a two-part, now apparently lost film. Here we see that Godzilla has seemingly forever been oddly associated with Kong since an early time. For example, the suit-maker for the 1938 Edo Kong flicks, Fuminori Ohashi, also participated in the construction of the first Godzilla suit. Also, Godzilla's familiar Japanese name "Gojira" translates to "*goril*la-whale" in English.

Then in 1948 a false Kong battled a dino-monster on an *Unknown Island*, directed by Jack Bernhard.[33] At a time when actors wore furred ape suits for popular films of the day, one such renown individual, Ray "Crash" Corrigan (1902–1976), starred in a movie scene anticipating another movie then as yet unconceived. Corrigan, wearing an ape suit in 100-degree heat, played the role of an extinct giant ground sloth *Megatherium* that battles a horned ceratosaur. Except this sloth should have been equipped with an enormous tail, lacking on the suit, instead enhancing the obvious gorilla-apish qualities instead of the prehistoric monster Corrigan hoped to somehow channel per the 1948 movie script. Those uninitiated today spying a famous movie still from this 1948 film would probably assume (incorrectly) that the suitmation actors were instead staging a rehearsal for a scene in 1963's *King Kong vs. Godzilla*, rather than *Unknown Island*—still relatively unknown today.

In fact, the very idea of a Godzilla wasn't even on the horizon in 1948, and wouldn't be until after the catalyzing 1952 re-release of RKO's *King Kong*. While of course there were other highly significant factors at large,[34] some would proclaim that Godzilla wouldn't have manifested in the 1954 movie *Gojira*, were it not, in part, for Kong's theatrical performances of the period. Afterward, while Godzilla ravaged Japan (and other giant paleo-monsters attacked western cities across the globe), another giant ape appeared, far exceeding RKO-Kong's modest proportions, ransacking London—1961's *Konga*, directed by John Lemont. Of the four giant gorilla-ape-monster suits of the 1960s, Konga's was the best. Yet of the movies of that period involving said suits, arguably *Konga* was not quite the worst. Although *Konga*'s giant ape didn't combat dino-monsters in film, he did in the tie-in comic book series published by Charlton.

Following an injection of carnivorous plant substance into the bloodstream of a chimpanzee, the chimp exponentially grows into an enormous, hypnotized gorilla who, in Kong-like fashion, snatches up a mad scientist. Eventually after going on a rampage, Konga shrinks back to normal size. But in Charlton comics' run, Konga has many more adventures, at least three of which involve tussles with reptilian-like creatures and dino-monsters. For instance, in *Konga* no. 2 (August 1961), Konga is transported to a lost world island still inhabited by dinosaurs; so, naturally he kills a tyrannosaur just before a volcano erupts. Then in *Konga* no. 3 (October 1961), now described as a hairy, primeval monster, Konga battles a coiling sea-monster thing, recalling Kong's 1933 battle with the Elasmosaur.

In 1963, Toho's *King Kong vs. Godzilla* reached American theaters, resulting in a number of copycat comic book stories showcasing giant gorillas battling dino-monsters, thus causing a wave of thematic excitement among monster-loving teens and dino-addicted boys well on their way to adolescence. For by 1965, we find several such tales invoking imagery borrowed, inspired by, or even suggesting Toho's classic 1962/63 film. In *The War That Time Forgot*, for example, a white-furred giant gorilla combats a spiky-looking dino-monster, driving it over a lost world–ish cliff setting (no. 124, "Terror in a Bottle," November 1965). Covers of books such as *Superman's Pal: Jimmy Olsen* no. 84 ("Jimmy Olsen's Monster Movie," April 1965), introducing Planet Krypton's fire-breathing "Flame Dragon" versus gorilla-ape "Titano," and *Konga* (no. 23, "The Creature of Uuang-Ni!," November 1965) featured dinosaur-fighting giant gorillas. And Charlton's *Fantastic Giants* (September 1966) implied a match-up between the reptilian dino-monster Gorgo and Konga (although this was a tease, because first issues of the *Gorgo* and *Konga* books were only reprinted therein without new material added concerning these filmic monsters).

Further in this context, we will defer addressing *King Kong vs. Godzilla*–themed iconography and films (both those made and never completed), although two additional Kong films should be mentioned here, as in each Kong tussled with prehistoric giant reptilians and dino-monsters.

Then, *The King Kong Show* appeared, a cartoon from Rankin/Bass, broadcasting 25 episodes from September 1966 through August 31, 1969. In one show, Kong battled a tyrannosaur on Mondo Island to rescue one of the human characters. According to John LeMay, a 1967 film, *King Kong Escapes* (Toho and Rankin/Bass, directed by Ishiro Honda), "is the result of both the cancellation of 'Operation Robinson Crusoe: Kong Kong vs. Ebirah' and the production of *The Kong Kong Show*.[35] In *King Kong Escapes*, the second-best Kong suit of the period (worn by Haruo Nakajima) fights, and kills, a vicious bounding, scaly suitmation "Gorosaurus"—not named in this film, but dubbed as such in 1968's

Figure 14: On the road through Kentucky, suddenly a towering titan of terror and tumult arises! (photograph by Kristen Dennis [2021], used with permission).

Destroy All Monsters (U.S. version post-direction by Peter Fernandez)—on Mondo Island. This dramatic footage was intended as a direct homage, an impressively orchestrated recreation of the famous Kong vs. tyrannosaur scene in the 1933 classic film.

Of the film itself coupled with the satisfying dino-monster suit (worn and enacted by Hiroshi Sekida), Mark F. Berry commented in 2002: "Maybe the most amazing thing … is that—karate-chopping tyrannosaur and all—it still offers more entertainment than the infamous 1976 version (i.e., *King Kong*) could ever hope for."[36] To me, though, despite Gorosaurus's casual tyrannosaurian appearance, its kicking antics harken back to a famous 1897 Charles R. Knight painted restoration of the smaller Cretaceous dinosaur Laelaps (also known as *Dryptosaurus*). True—dinosaur fans generally object to any filmic representation of Kong lacking confrontations with dinosaurs. And despite the producer's first name, a decade later, in Dino De Laurentiis's *King Kong* (1976, directed by John Guillerman) the only reptilian faced by Kong is a gigantic snake—a Freudian or religious reference, perhaps? Of the 1976 effort, Don Glut stated, "the new *King Kong* appears to be no more than an incredibly expensive Godzilla film and leaves no doubt that there is still only one *King Kong*."[37]

Well—there was one other filmic giant ape suit appearing during the 1960s, barely worth mentioning, but here goes—earning the award which many believe was earned for the worst giant dinosaurian-fighting ape-monster film of that decade. 1969's *The Mighty Gorga* (American General Pictures, directed by David L. Hewitt) was another movie in which the ape versus dinosaur/tyrannosaur theme played out once more. Of this, Mark F. Berry commented on the "sillier-looking than *Reptilicus*, this tyrann-eyesore takes the (cheese) cake. ... As bad as the dinosaur is, the Gorga suit is nearly its match ... with its immovable, googly-eyed visage ... it must be seen to be appreciated. Needless to say, the fight between the two is fairly hilarious—the picture's worst and funniest moment."[38]

Peter Jackson's visually stimulating *King Kong* (2005, Universal Studios) may have been intended as *the* climactic version of the Kong story, but in the end will only be regarded as an updated telling of the tale (not a sequel), albeit brimming with an incredible array of CGI enhancements. Here, Kong does battle dinosaurs—a particular genus of tyrannosaurs dubbed the "V. Rex" in the 2005 novelization written by Christopher Golden. The "V" stands either for "voracious" or "vicious." Why? As Golden explains: "They called the *Tyrannosaurus rex* the king of dinosaurs, but this thing was even bigger than the bones she'd seen in a museum. All she (i.e., Ann Darrow) could think was that this was the *T. rex's* evolutionary counterpart in this living hell."[39]

Unlike huge Konga before or perhaps the enormous Kong of Toho's Mondo Island, Kong in the 2005 film is a mere (yet powerfully built) 25 feet tall. In a fabulous fist-fighting feat that nearly defies credulity—even for those conditioned to having their disbelief suspended in such filmic fare—Kong acrobatically and simultaneously defeats not one, but *three* V. rexes! This protracted, approximately seven-minute battle, during which Ann resolves her allegiance to protector Kong, with exciting and magnificent heroine-always-saved-just-in-the-nick-of-time tempo, is definitely cool yet absolutely strains suspension of disbelief. At one point Ann even desperately clings onto the teeth of a V. Rex for dear life! (Incidentally, giant gorilla-ape "Titano" had previously battled *three* tyrannosaurs in *Superman* no. 147 [August 1961].)

Artist and writer Joe DeVito and Brad Strickland, with John Michlig, conceived unique visions of *Kong: King of Skull Island* (DH Press) in their sensational illustrated 2004 book. Departing rather from the 1933 tale, yet founded on essence of the original storyline, their enriched title is both a prequel and sequel to the RKO classic film, a story "inspired and based on the novel *King Kong* conceived by Edgar Wallace and Merian C. Cooper—novelization by Delos W. Lovelace."

For our purposes here, in a flashback sequence, Kong is put through his dino-monster fighting paces once again, battling a horde of feathered, velociraptor-like, possibly highly evolved Mesozoic survivals (i.e., "Deathrunners"), led by an intriguing, much larger 70-foot-long, 30-foot-tall bipedal, spiky "queen" known to the island natives as "Gaw." "For mysterious reasons at certain times some ... never stopped growing. It was believed this super-Deathrunner was their queen, necessary for the continuation of their species. Gaw was the worst of these malevolent monsters. She could direct the others in attacks."[40] Following a tumultuous battle, in true primeval style Kong defeats Gaw by ritually stabbing her in the neck with a *Triceratops* horn. Among the dino-monsters inhabiting Skull Island that Kong has fought throughout history, Gaw was clearly the most formidable. DeVito's many illustrations of Gaw, Deathrunners and Kong confronting these and a dino-monster tyrannosaur (smaller than Gaw) illuminate this fascinating novel.

DeVito and Strickland elaborated their examination of Skull Island's natural history, prehistory and delving further into the legend of Kong himself in another book, *King Kong of Skull Island* (2017). Here we learn of the Kong species, "unique among great apes," that has existed since the time of prehistoric man in Asia, some 15 to 20 million years ago. The authors wrote, "Though we have no skeleton of a *Gigantopithecus*, that of a Kong in some ways shows a closer kinship with humans than with other great apes,"[41] a statement which recalls that 19th-century debate between Owen and Huxley. Kongs have many jungle battles with dinosaurian monsters in this lavish, well-illustrated coffee-table volume as well.

Ironically, it eventually became known that Gorillas—the spark behind Ape Gigans and all the rest since—are not vicious animals, not threatening yet threatened, quite vulnerable in their diminishing habitat. Kong is iconic, in part because in his many symbolic manifestations he battled odzill that, like him, were out of deep Time. We *need* Kong because, despite his bestiality yet not our nemesis or foe, he is that who slays dreadful "dream dragons."[42]

Eventually, all things Kong *versus* as well as *and* Godzilla—their opposite-ness—will be discussed herein, including an infamous giant gorilla-ape suit of the 1960s. Note that in none of the examples discussed in this chapter was intent expressed on the part of the creators, respectively, for the monsters in question to portray or symbolize a means for maintaining or restoring planetary eco-balance—an ever growing 21st-century concern. It would seem that no pairing of fictitious giant prehistoric monsters could be of such disparate, distinctive origins, yet remain so inextricably intertwined in their tumultuous struggle to save Earth.

Figure 15: The myth of the violent, horrific gorilla was conveyed through 1939 advertising by the Ringling Bros. and Barnum & Bailey circuses.

CHAPTER FIVE

Conjuring Classic Godzilla

Godzilla, the monster, didn't suddenly materialize as the dinosaur we've come to know. It evolved into one, conceptually inspired by classic dinosaur restorations. Like a swirling solar nebula, the original Godzilla of 1954/1956 was a Frankensteinian synthesis of imaginations morphing, accreting bit by bit through a century and a half of scientific comprehension, artistic endeavor, fear of the past and anxiety over the future—involving numerous talented and visionary individuals internationally—that is, before condensing and suddenly via nuclear fusion shining brilliantly, bursting into stardom, transforming into that virtual, now ever so familiar cinematic conception. Let's use this opportunity to explore Godzilla's several saurian ingredients, formula, makeup—and the stories and supporting background. For Godzilla has a much longer history and ancestry than does Kong.

Several writers have pondered Godzilla's mystical dinosaurian origins before, including me, but perhaps not probing as far as we'll now delve into the monster's odzillaan beginnings. I have long been committed to the position that Godzilla ultimately should be regarded as a pseudo-dinosaur, and here's why ... (although some may disagree). Conceptually it all started with a dinosaur's tooth. And who gains credit for the synthesis that became the Godzilla suit?

The first veritable piece in the puzzle, or construct that is Godzilla, was noticed and described nearly two centuries ago, during the early 1820s in England. Before the *idea* of dinosaurs was familiar—with even their hallowed name still two decades henceforth—very unusual teeth and, soon after, fossils of an unusual, gigantic herbivorous reptile, were discovered in 1821 by physician/geologist Gideon Mantell (1790–1852)—an adept fossil collector—in Cretaceous deposits of the Tilgate Forest. Several teeth had also been found by his wife, Mary Ann Mantell (1795–1847), artist and manuscript collaborator with Gideon. Quarrymen had possibly shown Gideon samples of fossilized teeth as early as 1818. The collected fossils were shared with and inspected by other noted authorities—fossil experts of the time—resulting in contrasting opinions as to their nature.

One of these, the young science of paleontology's foremost genius of the age was renowned French anatomist Baron Georges Cuvier

Figure 16: Illustration by Mike Fredericks (2021) showing an image of the original Iguanodon tooth fossil as found by the Mantells, as well as a life restoration of the Iguanodon, stylized after Zdenek Burian's most famous representation of this dinosaur—the version which gave some life to Toho's Godzilla (used with permission).

(1769–1832), who in 1823 declared the provided specimens, including an upper incisor, belonged to a prehistoric rhinoceros. But this made little sense to Gideon Mantell because quarry deposits where the mysterious fossils were found predated those in which mammalia thrived. And it was not likely either that such fossils had eroded from younger, more recent strata, only to settle below in older geological beds, corresponding to an age of reptiles. Other bones were soon discovered in the Tilgate as well, including bones of the hand and also a strange horn-shaped fossil. A prehistoric rhinoceros, perhaps? After all, fossil elephant remains had been found in England too (although in younger rocks).

In 1824, William Clift, curator at the Hunterian Museum in London, resolved the matter after comparing the resemblance of the large fossil tooth to that of a modern iguana from Central America. So, it seemed that the Mantells' fossil reptile, previously unknown to science, may have been a gigantic iguana-like creature that (as indicated in early reconstructions and restorations persisting through the late 19th century, including those most influentially by Benjamin Waterhouse Hawkins) also sported a horn on its snout. Cuvier applauded this revised interpretation. Well, that "horn" was actually an elongated thumb bone, but this fact wasn't realized until the late 1870s after numerous additional specimens turned up in a Belgian coal mine.

Fossils assigned to this new prehistoric animal were described in Gideon Mantell's 1825 paper, "Notice on the *Iguanodon*, a newly-discovered Fossil Reptile, from the Sandstone of Tilgate Forest, in Sussex." The creature's name, *Iguanodon*, translates as "iguana tooth," as it had been christened by Gideon's colleague, William Conybeare (1787–1857), another pioneer in the study of fossil reptiles. *Iguanodon* was the second fossil creature to be later classified as a dinosaur (in 1842)—the first being the carnivorous *Megalosaurus*, another British genus (described by British geologist William Buckland [1784–1856] in 1822).

But the tale of this remarkable tooth requires further elaboration, as it illustrates how protean, Frankensteinian, restorations are, interpretations and imagery surrounding these old primeval creatures—dinosaurs—and hence the science fictional byproducts upon which were later founded over a century later. Such evolving imagery would thus appear genetic to the dino-monster breed. Gideon Mantell was preoccupied with the apparent and often exaggerated size of his dino-monster—at one time estimated to have been 200 feet long—proclaiming that "like Frankenstein, he was actually appalled at the being which rose from his meditations."[1] In 1832, he added another herbivorous dinosaur to his menagerie, the armored *Hylaeosaurus*, also discovered in the Tilgate Forest deposits. The trio, *Megalosaurus, Iguanodon* and *Hylaeosaurus*, were the first saurian fossils anointed by Richard Owen as dinosaurs on basis of their skeletal anatomy, in 1842.

Although *Iguanodon* enjoyed top-rung popularity among dinosaurs for several decades, by the early 20th century its reign had become eclipsed, yielding to other dino-monster genera, chiefly resurrected from North American strata and fossil trove. The long-necked dinosaurian breed, of which *Brontosaurus* and *Diplodocus* were emblematic, ascended to the crown of popularity by 1900 doubtless due to their staggering size and awesome museum displays, such as the *Brontosaurus* skeleton erected by 1905 in New York's American Museum of Natural History. Although popular lure of such huge sauropods was tremendous, resulting in several early 20th-century science fiction tales where they were cast as fearsome dino-monsters, they weren't nearly as ferocious as *Tyrannosaurus rex*—perhaps today recognized as *the* quintessential dinosaur, which eventually became symbolic of America's post-nuclear might. But by the time *T. rex* made its debut over a century ago, another herbivorous quadruped, the distinctively plated *Stegosaurus*, had already gone through several gestations in scientific restorations, including an early often neglected one which appears quite brontosaurian in nature. As we shall see through further elaboration, this particular trio—*Iguanodon, Stegosaurus*, and *Tyrannosaurus*—formed an artistic basis for Godzilla's later emergence in suitmation perfection, uncannily emerging from the darkest recesses of human imagination.

However, of these three real dinosaurs, Godzilla's genesis is mostly upright stance and carnivorous fanged tyrannosaur (~55 percent). Secondly, it is (~40 percent) stegosaurian—due to its vertebral osteoderms—and to least degree Godzilla is partly iguanodontian (~5 percent), due to its powerfully long arms. Before seeing how these dinosaurs became integrated into one via Frankensteinian artistic inspiration (from least to most percentage-wise, and in chronological order), let's briefly turn to *Stegosaurus* next (e.g., as we've already briefly introduced *Iguanodon's* morphological contributions).

Don Glut states in his *Dinosaurs: The Encyclopedia* (1997) that "*Stegosaurus* ... is one of (the) dinosaurian genera best known to the public. Since its discovery, this fascinating animal, largest of all known stegosaurians, has been figured in virtually every popular book about dinosaurs. Its striking appearance, usually depicted in life restorations bearing two alternating rows of roughly triangular dorsal plates and four paired tail spines ... is immediately recognized by almost anyone with even a modicum of knowledge about extinct animals. Much has been made in the popular press about this dinosaur's alleged "two brains," the brain in the head celebrated as "walnut-sized."[2]

Stegosaurus was described in 1877, from fossils discovered in Jurassic rocks of Colorado—the famous Morrison formation—as early as 1850. Confusion reigned for decades over the correct positioning of those plates (osteoderms) arranged along its vertebral column and several spikes situated at the tail end, resulting in several speculative life interpretations, the earliest of which (sort of a sauropod-stegosaur hybrid published in 1884) soon resulted in a fanciful, usually neglected 1886 sci-fi illustration (see Figure 6).

In fact, one of the earliest speculations, suggested in 1877 by paleontologist Othniel C. Marsh (1831–1899), was that *Stegosaurus's* plates were arranged laying more or less flattened over its back—like a turtle! (Hmm—so was Marsh conjuring a future, yet prehistoric variant of Gamera?) Eventually he and others realized that the plates were ordered in some configuration, vertically, along its spine. A single row of plates, a double-paired set of rows, or a double-alternating sequence of rows—the conventional mid–20th-century view? The discussion continued for most of a century thereafter, testing imaginations of scientists and artists alike. Their exact function, as well as orientation of the bony tail spikes—were they positioned upright, or angling defensively outward—was also debated. As later assured by Robert T. Bakker (b. 1945), these hallmark features were covered in integumentary (skin).

Following the sauropod-stegosaur hybrids of 1884 and 1886, quite a collection of innovative restorations and reconstructions emerged over the next few decades, the most peculiar of which was prepared by artist Frank Bond in 1899, as directed by Dr. W.C. Knight. Here, *Stegosaurus's* plates were placed in a procumbent position, and at the many sutures where plates met arose bony spikes. Bond's stegosaur, vaguely resembling Toho's later Anguirus although more sparsely spined, also was shown rising on its hind legs and tail in a tripodal position to feed on tree leaves. (See Figure 31 in Appendix.) As stated in 1914 by stegosaur specialist Charles Whitney Gilmore (1874–1945), however, "All of the evidence ... is opposed to such an arrangement of the armor."[3]

Other paleoartists weighed in with their interpretations of skeletal anatomy. Over a century the series of early evolving restorations has been detailed in a number of monographs and popular publications. (See Figure 32 in Appendix.) There are several species of *Stegosaurus* now known to science, but by 1987 consensus in the case of *S. stenops* would suggest that its 17 plates were arranged in an alternating single row, with four tail spikes slightly angled outward toward would-be attackers, based on Stephen Czerkas's 1986 sculptural restoration. By

1954, Godzilla's year of birth (as well as this author's), the most famous and most widely reproduced (i.e., in print), scientifically accurate, mid-century stegosaur restorations of the time had been made by paleo-artists Charles R. Knight (1874–1953), Rudolph F. Zallinger (1919–1995) and Zdenek Burian (1905–1981). Their works form a pinnacle of the golden age of paleoart. Theirs comprise likely source material inspiring the plate arrangement of jagged-looking, three parallel rows of fins along Godzilla's spine.

And then, long before the May 2000 debut of the Field Museum of Natural History's "Sue," came *Tyrannosaurus* from the Hell Creek formation of Montana. Barnum Brown (1873–1963) actually discovered remains of three *T .rexes*; the first in Wyoming in 1900 (which was shipped to London), another along the badlands of the Missouri River in Montana in 1902 (later shipped in 1940 to Pittsburgh's Carnegie Museum), and then again in 1907, which became the American Museum's world-famous specimen collected in 1908. But it was vertebrate paleontologist Henry Fairfield Osborn (1857–1935) who in 1905 named this dinosaur *Tyrannosaurus rex*, even before its bones were fully excavated. Osborn, not Brown, formally described the dinosaur in several papers, although guessing incorrectly about its proportionally small forelimbs (that he had assumed were more so larger and longer, "allosaur-sized" and three-fingered rather than only two-clawed, which was the actual case). Paleoartist Erwin S. Christman (1884–1921) made the earliest artistic reconstruction on paper of the dinosaur's skull in 1906—when the skull itself was placed on display in New York—for one of Osborn's descriptions.

A startling, partially complete skeleton (its actual forelimbs then still unknown to science) of North America's tyrant lizard king was disinterred from hard sandstone matrix, mounted and on display at New York's American Museum by 1915. Ever since, *T. rex* has been celebrated continuously in popular venues, such as in field discoverer's Barnum Brown's *Scientific American* 1915 article, "Tyrannosaurus, a Cretaceous Carnivorous Dinosaur: The Largest Flesh-eater That Ever Lived."[4] Here, Brown offered a "pen-picture" glimpse into the life of this new dino-monster, in which *T. rex* actively springs upon a group of meek duck-billed "Trachodons" (now known as *Anatotitan*) and then again upon a tyrannosaur challenger for the spoils. Decades later, the imagined tableau of tyrant-lizard stalking duckbills—founded upon skeletal restorations in the American Museum, as well as inspired by Knight's early paintings—was vividly captured in a magnificent 1938 painted restoration by Czech paleoartist Zdenek Burian.

Accordingly, the originally planned tyrannosaur exhibit in the

museum, as spectacularly modeled by Raymond L. Ditmars in miniature, was supposed to feature two tyrannosaur carnivores fighting over
a duck-bill carcass, but such intricate, fluid posing would have proven
unmanageable then, given considerations of improperly balanced metal
bracing rods. So instead, a conservative, static, influentially upright
pose was selected by Osborn, leaving museum patrons awestruck for
nearly eight decades, that is, until 1992 when its skeleton was reconfig-
ured in a more likely horizontal pose conforming to updated science
of the dinosaur renaissance era—later astoundingly captured in 1993's
Jurassic Park.

T. rex as the new colossal fossil became widely known enjoyed an
influential reign on Earth like never before, through the 20th century
in many representations celebrated throughout pop-culture, eventually supplanting popularity of all other dinosaur genera—although
often depicted in tandem with its favored primeval nemesis (e.g., thanks
to Charles R. Knight's artistic prowess), three-horned *Triceratops*.
Knight's first classic restoration of the *Tyrannosaurus* was completed in
1905. Afterward, between 1926 through 1930 he completed a brilliant
series of murals for the (then) Chicago Natural History Museum (now
the Field Museum of Natural History). One of these showcased another
Rex versus Tops Late Cretaceous showdown—perhaps his most famous
painting! (And my favorite dinosaur restoration of all time, defining what the dinosaur age is or should be all about!) Knight also made
other small sculptural and illustrated restorations of tyrannosaurs. He
envisioned the expanse of geological time, inspiring many other mid–
20th-century paleoartists, advertising artisans, sci-fi fantasy artists and
movie effects designers to portray this formidable dinosaur—*Tyrannosaurus*—sculpturally at life-size, mobilized through geared animatronics or stop-motion photography, projected at even larger than life
dimensions (as in RKO's *King Kong*). Yes, by the mid–1930s, *T. rex* had
even eclipsed the imposing *Brontosaurus* as America's most captivating
dinosaur.

As reflected in 2020 by paleontologist/paleoartist Mark P. Witton, "To some people, Knight *is* the early history of paleoart, or at least
the only history worth knowing."[5] Amazingly, Knight's signature misty,
atmospheric paleoart was resultant of a visionary artist who was legally
blind in one eye. Knight's career in restoring prehistoric, extinct animals to life began in 1894 when he completed a life portrayal of a Ter-
tiary mammal known as the *Elotherium* for American Museum scientist
Jacob Wortman. He then had a short apprenticeship under pioneer paleontologist Edward D. Cope (1840–1897). Knight's 1897 restorations
resulting from that fated collaboration, published in *Century Magazine*[6]

Figure 17: A 1930s newspaper advertisement for Messmore and Damon's World a Million Years Ago exhibit—a fascinating collection of animatronic dinosaurs and other prehistoric animals. Note their reliance on the image of a tyrannosaur, by then America's most popular dinosaur genus, further suggesting its huge, larger-than-life size in comparison to a diminutive looking Statue of Liberty.

remain influential and are still reprinted today in media. Paleontologist Henry Fairfield Osborn was so smitten with Knight's abilities that he soon had him commissioned to produce many paintings depicting fossils and prehistoric life accenting displays in the American Museum. Over the next half century, Knight's magnificent artwork showcasing prehistory and primeval organisms from the Precambrian Era to the dawn of modern man reached the public, globally, gracing books and magazine articles—several under his own authorship.

Soon there would be many more, ubiquitous artistic renditions of *T. rex*, many borrowed from Knight's extraordinary visions! Yet rather oddly, neither Knight's most famous and widely influential Field Museum Late Cretaceous portrait (featuring a dramatic showdown in the old, *old* west between *T. rex* and *Triceratops*), nor others he completed, were destined as *the* tyrannosaur depiction fated to lend its striking contours, power, mien and morphology to the Godzilla costume. That honor would fall to another paleoartist—Rudolph F. Zallinger— whose majestic *Tyrannosaurus* painted restoration decorated the hall of the Peabody Museum of Natural History in another distinctive mural celebrating the evolutionary *Age of Reptiles*.

By 1937, Russian-born Zallinger was an Artist in Residence at Yale University, but by 1942 was persuaded to paint a mural in the Mesozoic dinosaur wing of the Peabody Museum ... just a little something to "spruce it up."[7] This modest incentive resulted in an artistic odyssey—a "panorama of time"—through the Age of Reptiles, spanning 300 million years of geological time from the Devonian Period to the Late Cretaceous (moving from right to left in a viewer's perspective). Perhaps the single greatest dinosaur painting of all time! Collaborating with paleontologists Richard Swann Lull (1867–1957) and Alfred Romer (1894–1973), the magnificent 16-foot-high, 110-foot-long mural was completed in 1947. Within a decade, in 1957, Zallinger's "Rex" had proven so popular that a baggy-looking suitmation version of it was made to menace humans who had unwittingly ventured into Universal-International's sci-fi primeval film setting, *The Land Unknown* (directed by Virgil W. Vogel), starring Jock Mahoney. (Interestingly, Zallinger had occasion to watch *Jurassic Park* [1993], with John Ostrom [1928–2005], the Yale paleontologist who ushered in the dinosaur renaissance in 1969 through his description of the *Deinonychus* so anatomically similar to those filmic, avian-like raptors [i.e., *Velociraptor*] then invading popular culture. "Reportedly, [Zallinger] liked the film, especially when the lawyer got chomped on the toilet" by the bioengineered tyrannosaur.[8])

Ruling the Cretaceous end of the Yale mural stands Zallinger's majestic, lordly *Tyrannosaurus rex* looking fierce and commanding as it surveys all that led to its predominance over nature. This is the famous pot-bellied, upright-postured, grinning Rex, unconcerned by a nearby smoldering volcano, or even by its usual enemy—an adjacent *Triceratops*. This is the vision—*the primary dinosaur*—that seven years later infused itself into the Godzilla costume. How do we know this?

Figure 18: Illustration by Mike Fredericks (2021) prepared under the author's direction for this book, several weeks before release of *Godzilla vs. Kong*. Symbolically, this image represents the ape vs. reptile conflict, with a Tyrannosaurus (at right) borrowed from 1957's *The Land Unknown*, which in turn, in that movie, was a suitmation dino-monster inspired by Zallinger's mural *The Age of Reptiles*. (Undeniably, Toho's world-famous suitmation dino-monster of 1954 was far more impressive looking than the Zallinger derivative shown in this image.) Meanwhile, the giant apelike creature (at left), more akin to ourselves, is seen wielding a war club, ready to combat the relentlessly oncoming dino-monster using this tool. Which (if either) of the monsters will protect the puny humans shown at center, below? (Used with permission.)

We'll go there, but first, I've identified three pop-cultural, scientific interpretive ages of *Tyrannosaurus*. These are the "Savage Rex" era from 1902 to 1942, followed by the "Lordly Rex" era, which begins in 1947 upon completion of Zallinger's masterful mural, through 1975 (approximately when the "Renaissance Rex" era of tyrannosaurs begins, which need not concern us here). War-like, cold-blooded Savage Rex is championed by Knight's earlier restorations, while Lordly Rex has a different persona and subtext. As I stated in 2000, "Cold war era Lordly tyrannosaurids exude a self-confident, battle hardened, unthreatened demeanor, evidently secure in their role as Mesozoic overlords. If *T. rexes* of the atomic age, painted or sculpted by paleo-artists in America, seemed majestic, it was because they reflected America's might and authority as a world power."[9]

Zallinger's dominating tyrannosaur superseded Knight's in the aftermath of World War II, a war that ended ultimately, horrifically on an atomic battle field. And it is Zallinger's *T. rex*, not any of Knight's perceptions, that was assimilated into the Godzilla costume. In its Peabody manifestation, Zallinger's *T. rex* anachronistically, yet ironically as well, presides over both a fictionalized end-Cretaceous depiction and the stark reality of our Anthropocene. During production stages of *Gojira*, Toho artists knew of Zallinger's mural.

The embryonic idea of a Godzilla monster didn't necessarily begin as anything that could be described as dinosaurian, however, as Ed Godziszewski outlines in his informative 2004 article "The Making of Godzilla." Encouraged by successes of then recent American film releases in Japan, *King Kong*'s 1952 re-release and 1953's influential *The Beast from 20,000 Fathoms*, Producer Tomoyuki Tanaka's original concept for the as yet unnamed Godzilla concerned a monster that "was awakened by the shock of hydrogen bomb testing ... the energy from the bomb ... gave it very unique features."[10] Indeed! The working title for this proto-movie monster divulges its hallmark marine aspect, "The Monster That Came from the Sea." Furthermore, in wake of the *Lucky Dragon no. 5 (Daigo Fukuryu)*, disaster giant monsters were quickly becoming the zeitgeist in Japan, as others had similar ideas for a giant monster film as well. For writer Takeo Murata (1907–1994) and special effects master Eiji Tsuburaya (1901–1970) were also setting wheels in motion.

Murata recalled Tsuburaya's core idea. "He told me there would be this big creature like a whale or something in Tokyo Bay and that monster whale ended up landing in Tokyo and then going on a rampage."[11] A monster whale—like Moby Dick? Well, that would soon change. Meanwhile, Tanaka floated his idea to his boss, executive producer Iwao Mori, who in turn suggested he discuss the project's technical (e.g., special effects) feasibility with Tsuburaya (who had also alternatively conjured a gigantic octopus as the rampaging movie monster instead of a whale). According to John LeMay, however, Tanaka already conceived that the monster be a giant reptilian, a pseudo-dinosaur perhaps, not unlike Ray Harryhausen's Rhedosaurus—featured in *The Beast from 20,000 Fathoms*. Most significantly, though, because Tsuburaya believed the movie could be effectively produced, Mori green-lighted their joint monster film idea, now known as "Project G."

The whale/octopus concept would be shed as things got rolling—a screenplay was being written and while a visual monster design was in preparation. Enter Shigeru Kayama (1904–1975), skilled and award-winning science fiction writer, on May 12, 1954, who wrote Toho's original Godzilla detailed story treatment—later modified and embellished by Ishiro Honda and Murata. (Tanaka's story was expanded into a novel, *Monster Godzilla*, published in Japan in October 1954.) Although the term "dinosaur" doesn't appear in a recent translation of Kayama's story printed in Godziszewski's article, LeMay reports that "Kayama clearly intended Godzilla be a reptile."[12] Godziszewski further claims Kayama, who was knowledgeable in paleontology envisioned Godzilla as having been born in the Jurassic Period, "two million years ago." Clay model designer and sculptor Teizo Toshimitsu further opined that Godzilla as projected in the early script versions was "just a reptile, not a bit scary."[13]

Following Ray Bradbury's (1920–2012) and Ray Harryhausen's (1920–2013) bold lead, Godzilla's eventual celebrity at the pinnacle of dino-monster status seems to have stemmed from the marine pseudo-dinosaurian sector of imagined suspended disbelief. Bradbury's unnamed dino-monster in his classic 1951 short story "The Fog Horn" and Harryhausen's 1953 Rhedosaurus were both gigantic *marine* monsters, unknown to 1950s contemporary science. But another important intellectual resource was Kayama's own 1952 short story "Jira Monster," concerning a "bipedal lizard monster terrorizing primitive islanders."[14] If their fossils and anatomy were to be examined, such creatures would not have been classified as dinosaurs despite their scripted geological provenance. And yet Godzilla's imposing form was ultimately derived from three genera of dinosaurs—a trinity—initially interwoven by sculptor Teizo Toshimitsu into three test 40-centimeter-tall sculptural designs ... clay maquette prototypes. However, by this time the merging of three real dinosaurs had been assimilated into the general Godzilla design.

Zallinger's 1947 mural proved key inspiration for Godzilla's tyrannosaurian aspect or mien, reflecting approximately 55 percent of Toho's new dino-monster. We know this thanks to a photo appearing in a 13-minute documentary, Ed Godziszewski's informative "The Making of the Godzilla Suit," an extra featurette in ClassicMedia's 2006 two-disc DVD of *Gojira*. I had known for years of the apparent triad of dinosaurians assimilated into the Godzilla costume, including *Tyrannosaurus*. But one photo in this documentary left me gobsmacked!

Therein is a photo (seen twice in the video) showing Toho's team of artists and model designers, one of whom is pointing to a sheet that to me was immediately recognizable—the paleontological fold-out sheet poster from the September 7, 1953, issue of *Life* magazine displaying Zallinger's smaller preliminary version cartoon of his *Age of Reptiles* mural. The photo is a little smudgy (yet looking just a little clearer when shown a second time), but one may discern Rex, the king himself, on the printed page. This captured moment perhaps represents that mid-stream course-correction moment when Toho elected to give Godzilla his unique and distinctive yet dinosaurian genesis, relying on morphed classic visuals. And so, to a major extent *Life* gave life to Godzilla.

Another photo embedded in Godziszewski's featurette highlights an artist drawing an early version of a tyrannosaurian Godzilla monster (with rudimentary dorsal spines), while inspecting a 1941 gouache painting of *Iguanodon* by Zdenek Burian. Unquestionably, Godzilla's powerful long arms (and a preliminary, very low single row of dorsal spines), evident in the Toho artist drawing, were borrowed from Burian's restoration. Clearly, the actual proportionally puny arms of *T. rex* would have proved insufficient for a suit actor inside the Godzilla costume. Like *T. rex* (and other bipedal dinosaurs as then known to science), Burian's iconic *Iguanodon* restoration was also figured in a fully upright stance. Such an upright posture would have facilitated the suit actors' ambulatory motions on set inside the Godzilla costume.

So, is Godzilla a variant stegosaur? Yes, partly. We have Zallinger's 1947 tyrannosaur, Burian's 1941 *Iguanodon*, but which artist's stegosaur? And why those fins, or plates, along Godzilla's neck, back and tail? *Spinosaurus* (see Figure 30 in Appendix) was hardly known then, especially in popular accounts, but Toho's artists could have selected a continuous, thinner pelycosaur (e.g., *Dimetrodon* or *Edaphosaurus*) fin, instead of amalgamating deformed, jagged stegosaurian plates into their design. Well, although I don't know whether Toho artists ever even considered a pelycosaur fin, the concept of stegosaur plates just looked much cooler on the costume, I suppose (which is probably why the Burian-iguanodontian low dorsal spines were rejected in favor of stegosaurian "plates"). But, still, whose stegosaur restoration was most influential? I presume Knight's then world-famous Field Museum of Natural History stegosaur mural, yet Toho also had ready access to Zallinger's (i.e., in the Jurassic Period segment of his *Age of Reptiles* painting), and most likely, or perhaps, one of Burian's as well, that may have appeared in books available in Japan then—like that mysterious "children's dinosaur encyclopedia."

The subsequent story of the making of the first Godzilla suit has been well documented and chronicled, as a team of artists, sculptors and artisans set methodically to work. Next, the task shifted to clay modeling, executed by Teizo Toshimitsu. He sculpted three 16-inch-tall versions from which to choose. Godziszewski described each of these in 2004:

> Using the sketches from the art department as a guide, Toshimitsu sculpted several variations in clay. ... His first rendition closely resembled a Tyrannosaurus but with a large, wide head. At close to 40-centimeters tall, the model was covered in serpentine scales, suggesting the appearance of a sea creature. A second model reduced the size of the head and eliminated some of the serpentine features, adding more bulk to the lower torso to impart a more massive and ponderous look. This design, dubbed the "warty" Godzilla, used large rounded bumps for skin texture. The third and final model was dubbed the "alligator" Godzilla, using the same physical characteristics as the "warty" Godzilla, but substituting skin texture of small, linear bumps for skin detail. This design incorporated the horrible disfigurement that Godzilla must have experienced when exposed to the atomic bomb. It was this model which was approved by Tsuburaya, Honda, and Tanaka as Godzilla's final look.[15]

(Note—it was most likely not the atomic bomb but instead the far more potent thermonuclear hydrogen bomb, that Godzilla would have been exposed to or awakened by, and the weapon feared most by the world then.)

From there, the first prototype six-foot-tall, 220-pound Godzilla suit was constructed. This version proved too stiff for suit actor Haruo Nakajima (1929–2017) to walk in and trample on-stage model buildings, however, and so a second lighter costume was made that appeared in *Gojira.* Yet, a third version, further differing in detail, was made for Toho's movie sequel, *Godzilla Raids Again* (1955), as well as another suit for Godzilla's first dinosaurian foe therein—Anguirus, the pseudo ankylo-ceratops dino-monster.

Also endemic to the lineage of dino-monsters disinterred from stone was their metaphorical meaning in popular culture. More than the sum of its constituent anatomical parts and, beyond allegory or metaphor, Godzilla also *assimilated* the supposed *character* or persona of each of its three real dinosaurian progenitors—*Iguanodon, Stegosaurus, Tyrannosaurus rex*—further adding to his/its polysemous nature in films. So far we haven't fully summarized paleoartist Zdenek Burian's unwitting contribution to the Godzilla costume, his *Iguanodon* painting that became infused into components of the Godzilla costume. Burian may be the most prolific of paleoartists, completing over 500 prehistoric animal restorations and primeval landscapes (and thousands of others composed on differing themes).

Born in Moravia, his popular natural history and paleontological work reached western shores and Britain by the mid–1950s, especially through a series of magnificent volumes, penned by paleontologist Josef Augusta and translated into English, featuring Burian's art. Burian also completed considerable numbers of science fictional artwork to illustrate Czech editions of works by Jules Verne, Arthur Conan Doyle, Ray Bradbury, Vladimir Obruchev, Edgar Rice Burroughs and numerous other fantasy writers. While Godziszewski states the book used by Toho artists showing Burian's *Iguanodon* was a "children's prehistoric animal encyclopedia," without knowing the title it's possible that the book which they relied on was one of Augusta's prior prehistoric animal books published in Czech during the 1940s—predating *Gojira*. One wonders which other prehistoric animals appearing in that old encyclopedia might have been influential toward the Godzilla costume.

I am aware Burian's *Iguanodon* shown in ClassicMedia's featurette photo was completed in 1941, because this restoration also appeared later in a book which I've owned since 1959, Josef Augusta's *Prehistoric Animals* (1956), purchased by my dad in London. In Plate 27 therein it is signed and dated 1941, obviously thought to be the (then) "scientifically accurate" paradigm—updating considerably from Benjamin Waterhouse Hawkins's and Gideon Mantell's mid–19th-century interpretations. In fact, Burian's *Iguanodon* restoration was so influential that during the early 1960s the Marx Toy company based their small dinosaur figure of this genus on his 1941 painted restoration. (Marx also produced a 1950s pot-belly tyrannosaur toy figure founded on Zallinger's *Rex* as seen in the *Age of Reptiles* mural.) The life appearance of how *Iguanodon* moved was further updated during the 1980s by paleontologist David Norman, in collaboration with paleoartist John Sibbick.[16]

Dinosaurs (as well as dino-monsters) do take on a distinctive, sometimes anthropomorphic persona. Take the lure of familiar *Brontosaurus* for example, which reigned as popular culture's top dino-monster for decades during the early 20th century. Now recognized as a docile gentle giant, over a century ago its great size inspired not only awe, but also feelings of fear due to its human-imparted supposed belligerent demeanor (as in its raging cinematic struggles against terrible flesh-eating *Allosaurus*). *Brontosaurus's* status was eventually eclipsed by *Tyrannosaurus*—the Savage version—by the early 1930s, however, shy *Stegosaurus's* armor reservedly signified an impassive, defensive, leave-me-alone adaptational threat. Although *Stegosaurus* became fierce when provoked in battle (as in Disney's 1940 *Fantasia* versus a deadly tyrannosaur, or simply not quite so shy as in 1933's *King Kong*).[17]

Iguanodon is evolutionarily related to those duck-billed herbivorous dinosaurs of the Cretaceous Period, several of which were known as good-mother-lizards—the genus *Maiasaura* discovered in Montana during the late 1970s by paleontologist John R. Horner and Bob Makela. And so here don't we notice yet another of Godzilla's assimilated dinosaurian breeding coming to life—i.e., besides that ferocious, scary-fanged tyrannosaurian side? For, tapping into its ornithischian genetic inheritance embedded in classic paleoart we've discussed here, not unlike the peaceful, fossil, egg-laying *Maiasaura* that nurtured its young in nests, Godzilla was also a good model parent (mother or father, take your pick), as witnessed in 1967's *Son of Godzilla*.

So ... Godzilla physical appearance, as witnessed in this monster's first three films, is a wonderfully blended manifestation of three dinosaurs—*Tyrannosaurus, Iguanodon,* and *Stegosaurus*—thanks to classic, western influential paleoart conveying magnificent impressions of all three. And its on-screen persona became a Frankensteinian derivative assimilation of these three dinosaur genera; a fierce tyrannosaur, an armored stegosaurian automaton that shouldn't be provoked lest a radioactive bolt (rather than a sharply driven tail-spike) is issued, and later even a caring "good-father/mother-lizard" type of parental figure.

So, who does gain credit for the synthesis that became the original Godzilla suit(s)? Not a single individual, but certainly a congregate, staggered in time, of writers, artists, special effects masters and other talented imaginers—including early scientists—are all due credit. The Godzilla suit design arguably has a lengthy gestation, at first seemingly unpretentious and extending well into the 19th century. A Frankensteinian trinity of dino-monster progenitors (*Tyrannosaurus, Stegosaurus,* and *Iguanodon*) was woven into the Godzilla costume, fueled by merging classic paleo-artistic visions of three principal western artists—Zallinger, Burian, and presumably Knight.

But most importantly it was the stalwart Japanese team who in 1954 assimilated and completed the fusion—a reaction to the hydrogen bomb—creating that essential synthesis which became filmdom's favorite giant dino-monster star, initially coming to life in sculptor Toshimutsu's alligator design Godzilla clay sculpture. Ever since, Godzilla has taken on numerous manifestations and personas, stemming from visionary pioneering work, outlined here, that proved so influential then and now.

In 1963's Americanized *King Kong vs. Godzilla*, an "esteemed authority on prehistoric animals"—Dr. Johnson—claims that Godzilla is a kind of cross between a *Tyrannosaurus* and *Stegosaurus*, displaying a 1959 children's book. Well, although during his interview he showed an *Allosaurus* depiction rather than a *T. rex* to underscore his assertion, this is more or less what Toho's staff accomplished in their Godzilla suit design.

CHAPTER SIX

Let Them Fight

"Nature tends to compensate for diseases, to remove or encapsulate them, to incorporate them into the system in their own way."
—Frank Herbert, *Dune*, 1965, p. 275

"Earth will not destroy him. It is he who threatens to destroy the Earth."
—Loren Eiseley, *The Invisible Pyramid*, 1970, p. 154

Not unlike how rock 'n' roll music became the soundtrack of our lives (according to a Chicagoland radio station), science fiction and fantasy is our new modern mythology. And just as, according to an old adage, "all roads lead to Rome," in the case of Legendary's (and Warner Brothers') giant (anachronistically prehistoric) monsters, their 2010s series of dino-monster and daikaiju films inevitably, relentlessly led toward 2020's tumultuous showdown (rematch) between Kong and Godzilla battling to restore balance, plausibly construed as Gaian in nature to control the breach (nay, the scourge?) that is mankind.

Late in my college years (1975/76), I enthusiastically enrolled in several (upper level) undergraduate seminar/independent study courses with esteemed geochemist Robert M. Garrels (1916–1988), who, while instilling and broadening my horizons with knowledge concerning (low-temperature) geological cycling and occurrences of inorganic chemicals (silicates and carbonates, especially) on Earth, also stressed the importance of understanding biology—in context of natural processes. In retrospect, these edifying interactions took place while James Lovelock's Gaia theory, in which Darwinian evolutionary coupling of organisms with environment toward the persistence and continuation of life on Earth over a four-billion-year interval, was becoming more widely professed. For as Lovelock realized in 1968, "Living things must be regulating the composition of the atmosphere."[1]

During the late 1970s, though, Godzilla movies didn't reflect dawning Gaian philosophy. In fact, although environmental concerns are central to many Japanese daikaiju movies (think of Mothra's appearances for example), Toho never took up this cause quite to the extent that Legendary has in recent years, this coming at a time when climate change issues have become more in-your-face to western audiences. And if prehistoric monsters abound in modernity, then isn't this reflective of ecological chaos, or ecocide?

According to Legendary's Monsterverse chronicles, technological man has known generally of MUTOs (Massive Unidentified Terrestrial Organisms) since the 1940s, yet their significance has only recently come to light! A beginning of our end ... and of a new geological era?

Monsters Exist!

What do we mean by "balance"? Isn't this more or less what must be maintained—metaphorically—by a tightrope walker? With every step of the way the walker teeters slightly left, or right, gently up and down, his center of gravity oscillating in a controlled fashion, lest he tumble to the ground. Such a stunt is precarious indeed, yet if his system maintains proper balance, equilibrium throughout, minimizing oscillatory motion, he will cross the divide safely. Another relatively simple kind of balance is maintained by many home appliances, such as a thermostat—via negative feedbacks detected by a sensor, keeping indoor temperatures within a specified narrow range of comfort. In the tightrope walker example, the "thermostat" is the human brain intricately controlling nerves and sensation impulses to muscles on left, right, front and back so as to achieve proper balance.

But a planetary-scale ecosystem such as envisioned by James Lovelock and other Gaia theorists requires far more intricate (geophysiological and chemical) kinds of feedback controls to maintain balance sufficient for life's continuity—evolving plants and animals surviving in concert with changing environmental conditions through geological time. So detailed, in fact, that monster-movie directors and associated script writers can only scrape an ideological surface in projecting what *could* happen to our species—only one of many millions in the biosphere—if civilization inadvertently forces an oscillation too far in any direction, or, say, generates too broad a carbon footprint. After all, even the most godlike, majestic monsters are limited as to what they can do toward protecting our species within a ravaged planetary ecosystem.

In Legendary's (with Warner Brothers and Tencent Pictures)

Kong: Skull Island (2017), directed by Jordan Vogt-Roberts, we see Carl Sagan's vision of the ancient-aspect reptile versus prehistoric giant ape played out once more. In fact, ecological balance, as well as civilization's salvation, hinges upon this timeless proto-conflict. In his best filmic exertions of the past (1933, 1963, 1967, 2005), Kong battled dinosaurs, including Godzilla, for supremacy, while (only) protecting, in each case, a pretty girl. In the 2017 picture, Kong encounters a menacing subterranean species of two-limbed Skull Crawlers, designed to make your skin crawl, an "ancient species" of lizards that can grow to enormous

Figure 19: Photograph of Jason Croghan's 2019 miniature sculpture, exhibited at G-Fest, showing a Skull Crawler momentarily gaining an edge in subduing Kong before the former's ultimate demise in a Skull Island bay setting. Inspired by climactic scenes in the 2017 film *Kong: Skull Island*. (Used with permission. Sculpture and photograph by Jason Croghan.)

size—over 100 feet long! Kong's expanded role, herein, as chief ecological defender, is to keep this awful race at bay, confined to vents extending under the island. As in *Pacific Rim's* Kaiju, Earth's bowels once again spawn terrible organisms capable of threatening civilization.

Kong's eventful tale plays out in 1973 in an aftermath of the Vietnam War. With congressional facilitation, Project Monarch cryptozoologist John Randa (John Goodman) launches an expedition to an uncharted island, identified by satellite, in the South Pacific, "a land where God did not finish creation ... a place where myth and science meet." An ulterior motive, though, is to determine the underlying bedrock's consistency using seismic mapping techniques, triggering explosives. It turns out that the "bedrock is practically hollow." Channeling classic science fictional writings by Edgar Rice Burroughs (*At the Earth's Core*) and Jules Verne (*A Journey to the Center of the Earth*), while invoking the (beleaguered) Symmesian idea of a hollow planet may seem an improbable pseudo-scientific sleight of hand.[2] But this bit of fringe science will also matter in Legendary's 2019 film entry, to be discussed shortly (as well as the climactic 2021 film outlined in Chapter Nine).

The violent detonations summon an angry Kong comprehending what disturbance of the island's hellish under-layer will conjure—those Skull Crawlers, which as we later learn killed his parents decades before. In contrast, however, Kong is protective toward living things comprising the island's surficial indigenous ecosystem, which may be described as eco-pristine.[3] While Kong's island is downright prehistoric in the 1933 original, his 2017 insular, filmic habitat—as conveyed also in Legendary's 2017 graphic novel *Skull Island: The Birth of Kong—The Official Comic Prequel to the Major Motion Picture* (written by Arvid Nelson, illustrated by "Zid"[4])—features a combination of species either paying homage to the original film, or otherwise suggesting a primeval, naturally balanced ecosystem.

For instance, the Skull Crawlers were inspired from that 1933 stop-motion animated two-legged creature seen ascending along a thick vine from the "spider pit" ravine toward Jack Driscoll. And the 2017 spider with immensely long, spear-like legs recalls those spiders and other menacing arthropods that may, depending on which story you'd care to believe, have been filmed for the original, although now representing possibly lost footage.[5] In the 2017 movie, we see a fresh-looking *Triceratops* skull, recalling a scene in the 1932 *King Kong* novelization by Delos Lovelace,[6] while the deadly, winged flock certainly appears pterosaurian, pterodactyl-like. Furthermore, in the graphic novel we see a pack of furry-looking, velociraptor-like, canid Death Jackals threatening the expedition's team.

But as Keeper of Ecology, territorial Kong also defends other creatures on his island, while living in concert with a human tribe—the Iwi, who consider the gigantic ape a god. This is in accord with Merian C. Cooper's guiding principle: "if man does right by the monster, the monster will do right by man." Kong isn't just a big bipedal gorilla, but a "lumbering, lonely god," capable of unpredictably entering a "berserker rage, contrasting with an almost mythical ability to inspire religious awe."[7] In the graphic novel, however, Kong's defense of the Iwi seems more pronounced than in the 2017 movie. But the military/scientific invasion to Skull Island is wholly unwelcome. Colonel Packard (Samuel L. Jackson, an obsessed, Captain Ahab persona) wages war with Kong, vengefully igniting barrels of napalm in a jungle lake. Vietnam proved as unpopular and unsatisfying a war to Americans as would the death of intrepid, caring Kong at the hands of a military madman.

Certain creatures besides Kong whose relative size fluctuates between 100 and 200 feet tall, and the enormous spider, display gigantism on a startling scale. There is a gigantic octopus (harkening to another he fought in the 1963 *King Kong vs. Godzilla*) and a giant walking stick—nay, a walking log insect. Both the latter and a 45-foot-long horned water buffalo—a creature suggesting the Pleistocene—emerge from the mushy habitats thoroughly encrusted with a complex, symbiotic mini-ecology over their bodies. In the movie, Kong is shown carefully rescuing a buffalo that has been pinned to the ground under a downed helicopter.

Details within the informative, delightful graphic novel and movie differ considerably, both plot-wise and creepy critter–wise. While movie action takes principally in 1973, graphic novel Skull Island action events occur in 1995. Besides the exotic native species mentioned previously, the graphic novel also presents creatures such as the Sirenjaw, a crocodilian that grows to 65 feet long, and the heat-tolerant, 40-foot-long Magma Turtle whose young hatches from eggs in a volcanic regime. We see images in panels showing Kong battling the giant spiders, and Kong's ancestors at war with hellacious Skull Crawlers.

In the graphic novel, readers learn more of Kong—his history and planetary significance. For "Kong isn't just an animal. He is the King of that island, and every beast in it. As long as he lives, the rest of the world is safe." The Kong species inhabited the Skull Island, an Edenic "cradle of paradise" for millions of years, until the Skull Crawlers arrived through a "gateway to Hell." Ever since, the Kongs needed to hold them in check, commencing a "war of the gods ... of the Elder Earth."[8]

So, what would happen if Kong, the last of his kind, expired? Simple biology dictates that if you remove any species' natural competition then the surviving race will proliferate. Would mankind then face a Skull Crawler planetary invasion? After all, as clarified in Tim Lebbon's novelization of the movie script, Skull Crawlers have gills, meaning they could escape the island. Yet Lebbon also states that Kong had just confronted "the greatest monster known as Man."[9] One wonders: would those Skull Crawlers be any worse or less destructive than man if they got loose on the planet?

Bill Randa speculated that monsters "separated from evolution for millions of years" might exist in the underground world. He comes to a gruesome end, though, as described in the novel. Swallowed whole in the mouth of a Skull Crawler. Randa, who also had realized that the 1954 Castle Bravo nuclear detonations at Bikini weren't merely tests but actual attempts to destroy a giant then-unknown monster (which audiences realize was Godzilla), hauntingly asserted, "This planet does not belong to us." Man cannot own Gaia.

An important link exists between man and Kong—*our* monster—however, conveyed in one novelization passage where the female photographer, Weaver, senses a "connection across aeons of evolution."[10]

Kong's final, tumultuous battle with the "Big One" lizard, christened the Skull Devil, is no less than a symbolic match between fierce warriors of heaven and hell. For as Lebbon states, "Discovering King Kong must be like finding God."[11]

Finally, at the tail end of the film we are treated to the sight of ancient native engravings showing other MUTOs and elder gods; we recognize images of Rodan, Mothra, and Godzilla fighting Ghidorah. As we drift toward the sound of Godzilla's mighty roar, a Project Monarch cryptozoologist exclaims, "Kong is not the only king."

Titans!

Unnatural God

Paleobiologist Dr. Emma Russell (played by Vera Fermiga), refiner of the high-tech ORCA (possibly an apt abbreviation for Organism Radio Communicating Acoustics, per YouTube and the Gojipedia online website) bio-acoustical sonar device allowing rudimentary signaling with Titans, startlingly reveals her cards in *Godzilla: King of the Monsters* (Warner Brothers and Legendary Pictures, 2019, directed by Michael Dougherty), midway during the film when she insidiously proclaims:

> Humans have been the dominant species for thousands of years and look what's happened ... overpopulation, pollution, war. The mass extinction we

feared has already begun. And we are the cause. We are the *infection*. But like all living organisms, the Earth unleashed a *fever* to fight this infection. Its original and rightful rulers, the Titans. ... They are part of the Earth's natural defense system. A way to protect the planet. To maintain its *balance*. But if governments are allowed to contain them, destroy them, or use them for war, the human infection will only continue to spread.[12]

And Dr. Serizawa (whose father analogously in Toho's inaugural 1954 film *Gojira*) dies heroically, would also concur that Godzilla's role in nature is to maintain balance. The younger Serizawa (reprised by Ken Watanabe) has faith in Godzilla to spare man, even opposing use of the Oxygen Destroyer against the monster.

A metaphorical fever to fight the infection personified in man stands aligned with James Lovelock's belief: "If I am right that the glacial *cool* is the preferred state of Gaia, then the interglacials like the present one represent some temporary failure of regulation, a *fevered* state of the planet for the present ecosystem ... glacials are the normal state and the interglacials, like now, are the pathological one ... low carbon dioxide during the glacials can be explained by the presence of a larger or more efficient biota"[13] (my italics). Indeed, given that prolonged ice ages have endured for sometimes tens of millions of years, multiple times in primeval stages of Earth history, ice ages might appear to be a natural planetary state for persistence of life.

Paleoanthropologist Loren Eiseley so eloquently likened humans' ("world eaters") "precarious balance with nature" to a "planetary virus" in his acclaimed *The Invisible Pyramid*. For we are a species "of great technical skill, which has fallen out of balance with the natural world."[14] Echoing Loren Eiseley's 1970 views, paleontologist Peter Ward would further condemn man as the culprit driving the woeful pending and ongoing extinction events of the Anthropocene Epoch. Ward views mass (Medean) extinctions of Earth's past as a "rogue's gallery of planetary antibiotics ... and that anti–Gaian agent is humanity."[15]

Dr. Emma Russell, initially bargaining for a planetary "clean slate," is rather like a modern version of Dr. Frankenstein, using science (the ORCA) to revive monsters from the dead (or, rather, a near-death state of deep hibernation)—her most sinister deed being involvement with the freeing from Antarctic glacial containment of "Monster Zero," the 521-foot-tall Ghidorah (each of whose three heads have individualized personalities). As Greg Keyes, author of the masterful movie novelization (based upon the screenplay by Michael Dougherty and story by Max Borenstein), notes, "Rather than study fossils, she got to play with living creatures millions of years old."[16] Using the ORCA signaling device, Emma and Jonah rashly hope to control Monster Zero for their insidious purposes. In the 2019 film, man must determine which of the 17 paleo-monster Titans either hibernating or on the loose—the ancient Titans—are beneficial to man's existence—possibly like Godzilla—and which are opposed—like Monster Zero. As clarified in Keyes's engaging novelization, we learn considerable backstory concerning this dark enterprise.

Alan Jonah is an eco-terrorist, who at one time trafficked Titan DNA (a throwback idea to 2013's *Pacific Rim*), a dangerous enterprise. It would seem that DNA trafficking might pose dangers far worse in scale than "ordinary" drug trafficking. He's hell-bent on "restoring the natural order." How is that term defined? Jonah clarifies they're (meaning the Titans) "not the monsters. We are. The whole bloody race."[17] And Emma has become his ally. As long as Monster Zero is *contained* within the ice (during our ever-diminishing ice age interglacial), civilization would be relatively safe. But the deadly band of eco-terrorists resurrect this creature—the worst of the lot—an act metaphorically suggesting man's heedless, reckless disinterment of long-buried substances, harmful to the modern environment such as coal and toxic metals. Using ORCA (a symbolic Pandora's box), Emma also frees a fiery molten Rodan (whose bat-like wings in this film were inspired by Bela Lugosi-Dracula's cape) from an erupting volcano in Mexico.[18] By now, after being summoned by Ghidorah's mighty roar, technological kill-switches routinely fail; containment has been breached. Emma has forlornly lost control of the monsters. In fact, her daughter, Madison (Millie Bobby Brown), views her mother as the ultimate monster!

Keyes also explains how several of the Titans may have lived eons ago, including a theory on 355-foot-tall Godzilla's (whose size ranged between 300 to 600 feet tall in movie scenes) paleontological provenance. When the great Permian mass extinction occurred 252 million years ago (with an asteroid impact cited as cause in the novelization, which in reality was not the cause), 90 percent of life was extinguished. "It also left the radiation levels on the Earth's surface too meager to provide the massive reptile with enough sustenance. So he retreated to the depths, where he could subsist on radiation leaking up from the Earth's core. Other Titans had also gone into hibernation." A rival theory suggests that instead Titans "evolved underground, in huge hollows in the Earth, and return there in times of need." But when humans invented and detonated nuclear bombs, and began cruising in nuclear-powered submarines, "Godzilla and the other Titans began to take notice. There was 'food' up there again."[19]

The Permian catastrophe is further troubling because its occurrence would seem to run against Lovelock's Gaia thesis. As paleontologist George R. McGhee, Jr., stated in 2018, "To readers who subscribe to the Gaia hypothesis—that the geochemical and biochemical cycles of the Earth are buffered to protect and nurture life—the answer to (i.e., the cause of the Permian mass extinction) will be shocking ... and horrific."[20] So Medea prevails? After all, Ghidorah is a symbol of mass extinction. Or perhaps Ghidorah is more akin to the terrifying Egyptian serpent god of chaos—Apophis!

The Titans may seem unnatural in their invulnerability and capabilities, but Serizawa imagines that they must have evolved on a primordial Earth, a hellscape planet wracked by volcanic eruptions and radiation for billions of years, even before there was sustaining water or oxygen. "Only a few adapted, survived to live in an oxygen atmosphere, became immortal."[21] Yet Serizawa recognizes that Ghidorah is even less natural than the rest. Further, not unlike *Pacific Rim*'s kaiju, exogenous Ghidorah poses an extreme example of how life can be self-destructive.

Human overpopulation is suggested as a core problem of our plight. Emma naively perceives that Titans might co-exist with man, although only a substantially reduced human population, as in ancient days recorded in myth and legends. For only Titans can restore the planet to its natural order—perhaps a healthy, balanced Gaian system, achieved within a generation. In a sense radiation is the cure, for "wherever the Titans go, life follows—triggered by radiation. ... They are the only guarantee that life will carry on. But ... we must set them free."[22] Humans will not go extinct in her view, as man will then live in utopian harmony with the long-forgotten "first gods." But later, as millions suffer and die, while cities crumble in the Titans' wake, with entire ecosystems' destruction happening before her very eyes, globally, with Godzilla down for the count and Ghidorah reigning supreme—commanding ally Rodan and other Titans (other than Godzilla and Mothra)—Emma's mindset and motive drastically alters.

Mass-murderer Emma is having serious reservations about Ghidorah—that she's lost control of the monster revival situation. Instead, terrifyingly, Ghidorah has seized control—calling the shots! Peaceful coexistence would be impossible with Ghidorah in command.

Meanwhile, Serizawa is coming to a realization that this Monster Zero is not of this Earth—not of our natural order—that Godzilla was its ancient adversary, leading to Ghidorah's defeat and icy entombment eons ago. Ghidorah's likeness on an ancient cave painting underscores how this "dragon who fell from the stars—a hydra whose storms swallowed both *and* gods alike"[23] is leading the Titans toward mass extinction comparable to that of the Permian age 250 million years ago. Subservient Rodan's flight path seems to trigger volcanic eruptions.

Medean force Ghidorah's coming is likened to a massive storm center, of the heightened variety mankind faces during our Anthropocene age of climate change. And not unlike the *Pacific Rim*'s dino-daikaiju, it's "almost as if he's reshaping the planet to his own liking ... trying to tear the Earth's ecosystem back down to the bones and start over."[24] Rather than "restoring the planet," Ghidorah is using the monsters of legend to level extinctions, its impact on the weather (i.e., manifesting a Category 6 hurricane) is likened to that of a "nuclear winter."[25]

Once again, radiation reigns at the core. For, as Emma muses, "If the natural world had been in balance, the MUTOs wouldn't have emerged—and if they had they wouldn't have had piles of nuclear materials to feast upon. Without them, Godzilla would never have come out of his deep retreat to fight them."[26] Later, like Frankenstein's monster, Godzilla must be resurrected from near-dead in order to combat its ancient nemesis, Ghidorah, once more. Fortunately, Ghidorah *doesn't* control another key kaiju ... 800-foot wingspan Mothra, Gaia-incarnate, a "divine creature who's beautiful and elegant, but also deadly and precise!"[27] In Mothra there is Hope—hope for the human race in spite of everything.

Serizawa believes that in Godzilla there is redemption for man. He also believes that Godzilla proves Emma's concept of a planetary clean slate with co-existence possibility. Emma's fundamental mistake was in presuming that Ghidorah is a god of the natural order when instead it was a false Titan, a monster from somewhere beyond, an unnatural adversary to the Titan's "first gods." Relics witnessed in the sunken Atlantean city underscore man's former coexistence with *our* Titans in ancient history. Here the Hollow Earth, lost, sunken city connection is intriguing. In a way this is a throwback to New Age, Atlantean, Gaian-Earth consciousness enlightenment so prevalent in late 20th-century popular culture. In an effort to suspend disbelief, Legendary is more so relying on pseudo-science rather than (mainstream paleontological) science.

Another unlikely connection concerns the ORCA device. Not unlike the oxygen destroyer, the ORCA is also a potent weapon—in its original configuration combining bioacoustical signals from several Titans into a recognizable frequency, in order to allegedly control resurrected Titans. However, if improperly wielded in battle or should it fall into misguided hands, tuning into the wrong frequencies could inadvertently result in a thousand San Franciscos (this a reference to the havoc and destruction witnessed in Legendary's 2014 film). Ultimately the ORCA signal responsible for 2019's far worse global cataclysm—capped by Ghidorah's usurpation of Titans' control—was modulated between benevolent Godzilla and ourselves! ORCA's bio-acoustical signal is composed from human and Godzillean characteristics—man and monster melded anthropomorphically! A reversion to ancient times. The fallacy is that Titans could truly be controlled by any device, even a cleverly devised ORCA.

Amidst the charred battlefield, a demolished Boston appears as a "volcanic wasteland millions of years ago."[28] Protagonist Dr. Mark Russell (Emma's estranged husband) imagines himself as a "little rat-like primate ancestor watching the gods fighting it out, desperate to find a place to hide."[29] For a while Mothra and Godzilla duke it out against Ghidorah and a flaming, magmatic Rodan in the fiery shambles of Boston. Godzilla—juiced on radiation while embodying the nuclear theme once more, battles Ghidorah for supremacy among the Titans, yet faltering—requires a charging boost of potent pixie dust from dying Mothra. When a resurgent Godzilla goes supernova, the resulting thermonuclear explosion ends Ghidorah's dark reign on Earth (or at least until their next tumultuous match).

Kong and the Skull Crawlers sensed Ghidorah's call to action as well. But while the Skull Crawlers were waking and Skull Island's winged pterodactyl-like Leafwings make an appearance in Boston, Kong is uncaring, not thinking globally enough. Instead, Kong, holding his position, remains focused on holding those nasty subterranean Skull Crawlers at bay, within their abyssal domain so as not to disturb insular ecology. As Keyes remarks, "Change was not good. ... Let them (i.e., the other Titans) stay away. ... He (i.e., Kong) did not care about their places, their islands."[30] Kong's fated meeting with the more primordial species, Godzilla (perhaps afflicted with a *global* sense of eco-consciousness), must wait until another day, another battlefield. At this juncture, oddly by comparison, Kong seems strictly territorial and perhaps more nationalistic than Godzilla.

Gaian Fallacy

Legendary's 21st-century paleo-monsters (Earth's Titans, *sans* Ghidorah) increasingly have become symbolic and metaphorical for a balanced Gaian, nurturing global ecology, rather than—traditionally as in mid–20th-century Japanese films—radiation spewed by the H-bomb. And yet isn't this an unachievable holistic, utopian ideal ... one we can never return to, one that never existed? Is the Gaia hypothesis mere pseudo-science, a strange religion? Does Earth deliberately or uncontrollably poison her children? If we consider a very real natural laboratory—that aforementioned end-Permian catastrophe, the biggest of the Big Five mass extinctions—we may glean further insights.

To underscore his point concerning this mass-extinction event of unequalled enormity and magnitude, paleontologist George R. McGhee, Jr., without referring to Ward's Medea hypothesis yet in utter contrast to (or as a foil for) Lovelock's Gaia, instead invokes the Hindu god Shiva, "a god that both creates and destroys entire worlds."[31] Most readers know by now of the six-mile diameter bolide (asteroid, or possibly slightly larger comet) which impacted Earth 66 million years ago, wiping out the (non-avian) dinosaurs and many other species. But far fewer may know that the most intense and destructive mass extinction suffered on Earth occurred 186 million years earlier, at the end of the Paleozoic Era. And the causal mechanism of the end-Permian mega-catastrophe stems from within the Earth system, not incoming from outer space (like Ghidorah)!

For brevity I shall refrain from in-depth discussion of how the end-Permian event (also briefly mentioned in Keyes's 2019 novel[32]) began or what its dire consequences were, yet such details may be found in McGhee's 2018 fascinating (if not sobering) book *Carbonifer-ous Giants and Mass Extinction.* Essentially, though, over the course of 200,000 years (a relatively short geological interval of time), during waning millennia of the Permian Period, a hot magmatic super-plume ascended through Earth's mantle, blasting through what is now a portion of Siberia, erupting a volume of as much as 16 million cubic kilometers of basaltic rock. Warmed oceans were acidified, land was baking hot—lethal in tropical zones—global warming became a long-term consequence persisting for five million years, atmospheric ozone layer destruction—caused by extreme volcanic chemical emissions—permitted damaging ultraviolet to penetrate to Earth's surface, further polluted by toxic metals in extruded mantle rock. Anoxia (lack of oxygen) within oceans became widespread and biologically overwhelming. Only a few "disaster species" managed to survive the ordeal in marine and terrestrial settings: most species thriving in the prior, distinctive Paleozoic world on land and at sea suffered extinction.

So, is it possible that Ghidorah *may* represent a non–Gaian, Medean force? Perhaps instead, this monster *is* more akin to McGhee's suggested god Shiva—linked as well to the end-Permian mass extinction. Peter Ward introduced the Medea hypothesis in context of terrestrial biological causal factors triggering life's self-poisoning events, leading to several major mass extinctions in prehistory. Whether the cause of an era-terminating catastrophe reaches Earth's surface either from far below or above, it doesn't really matter which mythical god is tied to the event. But a magmatic superplume erupting from Earth's mantle—the root causal mechanism for the end-Permian cataclysm—is not clearly Medean per Ward, as it does not directly in itself demonstrate life self-exterminating via poisoning of the ocean-atmosphere system. Instead, there was an inorganic, geological trigger, perhaps a Gaian-neutral one, analogous to the dinosaur-killing asteroid event 66 million years ago. However, both catastrophic events triggered, or cascaded toward worldwide, *biogenic* self-destructive events.[33] So perhaps not unlike *Pacific Rim*'s dino-derived Kaiju from a mysterious Anteverse, Ghidorah's overwhelming impact on Earth may well represent a reincarnation of Shiva—destroyer of worlds, taking place not in primeval times but in modernity. Another cinematic, metaphorical warning.

Director Michael Dougherty did embrace godlike qualities of the monsters in this film. As he stated in *Cinefex 165*, "To me, these creatures are gorgeous in the same way that whales, lions and elephants are gorgeous. It was important to me that all the creatures were captured in a way that made you feel a sense of awe and wonder and not just terror or thrills. ... I didn't see guys in rubber suits. I saw ancient gods."[34]

At this juncture we have Mothra and Godzilla as presumptive Gaian-Earth symbols. Formidable Ghidorah is to Godzilla what Skull Crawlers are to Kong! Kong as Keeper of Ecology would also appear in tune with a balanced habitat continuing the primordial, protracted war between mammals and reptiles, so intricately linked to our own evolution and current dominance in the Anthropocene through an Era-ending invader (asteroid or comet) from space. Meanwhile (anti–Gaian) Ghidorah—the unnatural, alien god—represents everything antagonistic to achieving harmonious balance on Earth, a condition Kong is only able to maintain on a delimited (insular) scale, while Godzilla's scope outreaches globally. (Legendary's Rodan would seem a mercenary combatant.) It would seem that Kong and Godzilla must make amends if Ghidorah's invasion would be forever nullified.[35]

Loren Eiseley noted, "When man becomes greater than nature, nature, which gave him birth, will respond."[36] And so Carl Sagan's timeless metaphorical war, Mammal versus Great Reptile, is destined to recommence on its grandest scale. We are most closely linked to Kong in the evolutionary scale, yet Godzilla seems more attuned to civilization's welfare. A dilemma?

CHAPTER SEVEN

Godzilla and Kong
Primal Struggle

"The monsters among us are aberrations. They're a reflection of who we are. What we do with them and how we decide to root them out is also a reflection of who we are."
—Actress Jayne Atkinson as U.S. Attorney General Ruth Martin in CW's *Clarice*, first aired on Feb. 25, 2021 (season 1: episode 3)

Monstrous Metaphor

Although Merian C. Cooper stated that his super-monster of 1933, King Kong, lacked hidden messaging or symbolism, others of scholarly persuasion might disagree, as much has been written and read into the metaphorical meaning of the giant prehistoric ape. Clearly, Kong's tragic exploitation is at the heart of that story. With 1954's Godzilla entry, however, there's less controversy: this fire-breathing dino-draconian is a tank-stomping metaphor of modernity—technology—gone wrong. And so now a *merging* of these two most famous monsters creates a psychical coalescence of ideas transcending filmic conformity. Here, we're confronted by a collision of the primordial with the civilized world. Yeah—Kong versus Godzilla—*the* title bout at last, the world series of monster fights. It just doesn't get any bigger or better than this!

As recounted by Donald Glut, "Cooper has disavowed ... 'insights' into his film; he was not the kind of person to consider injecting such messages into his productions."[1] And what an array of messaging has been read into this masterful film by intellectuals—some ideas of more reasonable merit, given Cooper's stated conscious inclinations—others straying into the prurient, subliminal realm! For as stated by Steve Vert-lieb, "Great art ... lends itself to philosophical interpretation."[2] (Delving deeply into the subconscious where reside "monsters from the Id" may be precarious.) Assimilating such references under contemporary cir-cumstances, ideology surrounding Kong as Christ figure, or as a tragic, racially repressed "primitive" coerced into "civilized" servitude, and other suggestions per Vertlieb and Bruce M. Tyler does seem founded—notions that never entered my once-six-year-old mind when I first fell in love with this movie monster.[3]

As Paul A. Woods stated in *King Kong Cometh*, Kong "remains an archetypical monster of the imagination to anyone who ... first encoun-tered him at a sufficiently impressionable age. ... Kong ... possesses a subterranean growl that rises from the pit of his stomach like a volcano ejecting lava from Hell."[4] Okay, but now what of Godzilla?

Whereas Kong's gestation stems from a prehistoric setting, explored by the plundering white man during an extended age of geo-graphical discovery and colonization, Godzilla's core meaning is tied to modern mad-scientists on the fringe of knowledge run amuck, cursed by a predictable inability to control their powerful inventions. Over decades and through many movie appearances, Godzilla's true or singu-lar meaning has become polysemous, ever difficult to distill. As noted by John J. Pierce in *The Official Godzilla Compendium*, "Godzilla has infinite powers of regeneration. But that regeneration can also be con-sidered a metaphor. ... Godzilla is, and always shall be, a monster for all seasons."[5]

But in its original feature, 1954's *Gojira* (directed by Ishiro Honda), Americanized into 1956's *Godzilla, King of the Monsters* (directed by Terry Morse), existential threats posed by hydrogen-bomb testing in the Pacific region due to proliferation of nuclear weapons were most urgent. Instead, Godzilla is a revisitation of the Frankenstein myth, replete with anti-nuclear messaging. Steve Ryfle noted in 1998 that "Godzilla ... is a paradox ... a horrible embodiment of the Bomb that created him, and yet a pitiable victim of it."[6] Ishiro Honda "took the characteristics of an atomic bomb and applied them to Godzilla."[7] Or, as star actor Akira Takarada ("Ogata" in *Gojira*) summed, "I consid-ered the monster a warning from the Earth. ... He came ashore to bring a message to humanity. ... Godzilla is a problem made by humankind, and it has to be killed mainly because of man's selfishness."[8]

And for our present purposes of examining Kong alongside Godzilla, the nuclear metaphor remains that which may be differenti-ated from other, later allegorical characterizations evolving after Toho's 1962/63 release of *King Kong vs. Godzilla*. Ryfle continues on *Gojira*,

"Godzilla's hell-born wrath represents more than one specific anxiety in the modern age—it is the embodiment of the destruction, disaster, anarchy, and death that man unleashes when he foolishly unlocks the forbidden secrets of nature, probes the frightening reaches of technology and science, and worst of all, allows his greed and thirst for power to erupt in war."[9]

It is rather common knowledge that *Gojira*'s filmic ancestry extends to both *King Kong* (1933) and perhaps the more closely, thematically linked *Beast from 20,000 Fathoms* (Warner Brothers, directed by Eugene Lourie, 1953). Differences are more apparent between *Gojira* and *Kong*, yet there's a significant distinction as well between Toho's entry in the giant monster genre and *Beast*. The "Beast" Rhedosaurus was inspired by Ray Bradbury's 1951 short story of the same title—later retitled "The Foghorn," the tale of a lost, last dinosaurian attracted by the lonely wail of a foghorn, having nothing to do with nuclear anxiety. In *Gojira*, resulting from Shigeru Kayama's story treatment titled "The Giant Monster from 20,000 Miles Under the Sea," (eventually novelized in Japanese in July 1955), the titular monster is summoned by the test blasting of hydrogen bombs in the Pacific, causing it to become radioactive—quite a different sort of yarn.

Yet in a predecessor 1953 American film, the Rhedosaurus's radioactivity—which must have been present due to Operation Experiment's recent detonation of a hydrogen bomb in the Arctic—is subdued, hardly mentioned, or simply assumed. The primary danger posed by this long dormant saurian, stop-motion animated by Ray Harryhausen, is *not* radioactivity, but instead a germ released from its blood, causing a virulent contagion in New York City. Partly, this may be because, at the time, Americans were (politically) conditioned to feel more content—less anxious—with the prospect of radiation than were Japanese. After all, we dealt it. In certain respects, the 1956 American Godzilla version, in which *Gojira*'s original and vibrant anti-nuclear theme seems relatively suppressed, became a gateway for other giant monster films of the period, perhaps more so than any predecessor film.[10]

A new kind of super monster was needed for Toho's "Project G," but inspiration for its design was not quite as forthcoming as it was in the case of Cooper's Kong—partly inspired by W. Douglas Burden's bring-'em-back-alive stunt to the Bronx Zoo, where Indonesian Komodo dragons were first displayed during the early 1920s.[11] Curiously, it was in early 1954 when *Gojira* producer Tomoyuki Tanaka's (cancelled) movie project *In the Shadow of Honor*, intended for filming in Indonesia, was aborted that he arrived at the idea for a giant monster movie in a fevered fit of "clutch creativity."[12]

Godzilla's nuclear affliction is prevalent through the monster's first two movies (1954 and 1956), and 1955's *Godzilla Raids Again* (directed by Motoyoski Oda)—Americanized as *Gigantis the Fire Monster* (directed by Hugo Grimaldi, 1959), in which a second Godzilla defeats a giant saurian known as Anguirus. Here the anti-nuclear theme is subdued, yet still remains evident as both monsters were awakened from Mesozoic marine sediments by hydrogen-bomb testing. The original Godzilla had been irrevocably destroyed by a weapon even more terrible, ecologically destructive and potentially harmful to Gaia than the hydrogen bomb, the Oxygen Destroyer. By (1962 and) 1963—when *King Kong vs. Godzilla* was released in America—the radioactive theme was even further diminished yet not entirely abandoned.

Much has been written about the making of the Godzilla suits, so distinctive from any other dino-monster. But his/its primary assault weapon—the radioactive fiery ray issued from its maw (and other orifices, given Toho's *Shin Godzilla*), as conceived in Kayama's original story—activated by those flashing stegosaurian osteoderms along its spine seems taken for granted by audiences. As documented by Godziszewski, "the atomic breath was considered to be one of the critical elements of the project ... something which set it apart from similar films." Ishiro Honda stated, "What was most special was we could visualize radiation itself. By opening his mouth and simply exhaling, Godzilla could vaporize an entire building ... radiation ... is capable of that. Godzilla exhaling radioactive flame was not unnatural in the context of this film."[13]

Here is dual symbolism, because Godzilla's powerful death ray is a weaponized manifestation of what physicists were experimenting with decades earlier via their discovery of radioactivity, cabalistic investigations which led to the hydrogen bomb. The dino-monster's brilliant ray both denotes (and is a result of) modern nuclear weaponry. John J. Pierce adds, Godzilla is "supposed to be a dinosaur. ... But the fishermen of Odo Island identify him with a legendary monster of the past that could be appeased only by virgin sacrifices—which puts him (*also*) squarely in the tradition of the *fire-breathing* dragons of folklore"[14] (my italics).

Interestingly, author Mary Roach noted a suggestion by Stephen Secor in her 2013 book *Gulp* that pythons may pose a reason underlying the myth of fire-breathing dragons. Pythons (i.e., serpents) loaded with hydrogen gas due to decompositional reactions occurring within, when stepped upon near a campfire in olden or prehistoric times *might* have issued an unexpected sudden burst of flame. After all, "The oldest stories of fire-breathing dragons come from Africa and south China: where the giant snakes are."[15] Godzilla uses this ray to finish off Anguirus in their 1955/59 Osaka battle.

Had the Anguirus won in that contest, there may have been no extended Godzilla franchise. Because analogously to Kong, Gigantis was last of the Godzillean species. Elsewhere I have shown that another primordial battle, so visually captivating and influential to (us) dino-philes ever since the early days of paleoartist Charles R. Knight, is that of a dino-monster tyrannosaur versus horned, bull-dinosaur (e.g., *Triceratops*). Chimeric Anguirus may be considered, in part, besides ankylosaur, a sort of pseudo-ceratopsian (e.g., horned) dinosaurian. The fact that the titanic Gigantis vs. Anguirus on-screen tussle is considered one of Toho's lesser efforts (but not by this author[16]) is indicative that the apes versus dinosaurs/saurian theme seems even more provocative and primal than sensational Rex vs. Tops imagery and imagetext.

"After her, Malone. She's invaluable."

In a very large sense, then, we have stop-motion animator and artist Willis O'Brien partially to thank for Godzilla's continued heritage, despite his misgivings concerning two motion pictures of the early 1960s, produced during his declining years—which he had some pull on, at least for a time. These two latter major movie productions were the 1960 remake of *The Lost World* (produced by Irwin Allen), and another which never bore fruit—"King Kong vs. Frankenstein"—the title that ultimately (disappointingly, many might claim) led to Toho's blockbuster, *King Kong vs. Godzilla*. Was O'Brien striving for sensationalism, while also groping to relive past filmic glory? Well, at least some stop-motion animation—O'Brien's expertise—*was* employed in the latter film, even though he had nothing to do with it.

O'Brien had by then recently collaborated with Ray Harryhausen and Irwin Allen on a memorable stop-motion animated prehistoric life segment of a film titled *The Animal World* (Warner Brothers, directed by Irwin Allen, 1956). Later, by 1960, Allen was plotting to redo a modernized version of *The Lost World*, a film for which he consulted with O'Brien on how to accomplish special effects of live dinosaurs—and for which, to the latter's chagrin, living lizards and reptiles were enlisted instead of relying on sculptural models and articulated puppets. It seems that Allen, who'd remained patient before during production of *The Animal World's* dinosaur scenes, realized it would take much too long for animators to create scenes required for *The Lost World* if stop-motion animation were used.

I'm not ashamed to say that 1960's *The Lost World* is one of my very favorite dino-monster movies, although it clearly has its many flaws. Critical response was varied, but the review I enjoyed most appeared in London's *Daily Telegraph*, referring to actress Vitina Marcus, who played the lost plateau's attractive native girl. This whimsical review, reprinted in Steve Archer's *Willis O'Brien: Special Effects Genius*, read, "The *dialogue* is often a delight. I liked 'After her, Malone, she's invaluable!' referring to a nubile native girl with a cleverly-cut sarong and a very pleasing pair of legs. 'How horrid, eaten alive!' also makes its mark."[17]

Herbert M. Dawley, O'Brien's former film collaborator during the 1910s, and eventual nemesis concerning patent rights to stop-motion animation photography, perhaps had the last laugh. Dawley recorded in his diary on June 21, 1969, while watching television that "at 11:30 I tuned in 'The Lost World.' It brought back memories of my dispute with the Producers over my patents. It is a tawdry production." It may be that Dawley didn't realize how limited was O'Brien's "Effects Technician" role on this 1960 movie. O'Brien was relegated to merely rendering "preliminary designs for what would were to be stop-motion dinosaurs before Irwin Allen elected to use real lizards" instead.[18]

Although O'Brien would have much preferred performing his trademark stop-motion wizardry before the cameras for 1960s *The Lost World*, and also for the ensuing 1962 Japanese entry that eventually unfolded from his "Frankenstein" idea, such animation was clearly precluded. As Eiji Tsuburaya stated in 1962, "In America, films like *The Lost World* (1925) and *King Kong* were made with very advanced special effects technique. In Japan, we couldn't do this—for one thing, there was no one in Japan who was experienced in model animation. Our film (*Gojira*) was a big experiment. So to cover up all its shortcomings, we needed to have a better *story*. That was the idea. So Kayama carried a big responsibility."[19]

Of course, Kayama admirably succeeded. Interestingly, as noted by Ed Godziszewski, Kayama—who evidently was a paleo-phile—even established an intriguing, if not incriminating, link between man and daikaiju creature in stating Godzilla's supposed "Jurassic" origins were "only" two million years ago in prehistory. "That means Kayama intentionally overlapped Godzilla's birth with that of man's ancestors. He wanted to link them together."[20]

And so, coupled with Haruo Nakajima's heroic performances, Toho's special effects staff headed expertly by Tsuburaya—who thought a stop-motion animated version of *Gojira* would require *seven years* of filming—went on to perfect the art of suitmation![21] "While Eiji

Tsuburaya may have regretted his inability to use stop motion effects, the residual benefit of this situation was that it forced him to use his ingenuity and creativity to make the impossible possible, leading to many innovations."[22]

"King Kong vs. Frankenstein" was dabbled-in and dangled toward prospective financial backers and producers, before fatefully twisting on the vine. While O'Brien drew many sketches of his bellicose beasts, the "Kong vs. Frankenstein" idea morphed into "King Kong vs. Prometheus." And when Toho provided necessary backing, "Prometheus" instead manifested into their Godzilla.

O'Brien's wife, Darlyne, recalled the story outline in a ~1979 interview with Kevin Brownlow, stating of the Frankenstein film,

> This picture ended in San Francisco, and ... was gonna [*sic*] have King Kong riding a cable car, but (Willis) didn't get the illustration finished.... That was ... when he had the idea of "King Kong vs. Frankenstein." He did (have a story), but it ended up as *King Kong vs Godzilla*. He didn't have anything to do with it, but it was supposed to be what they were promoting, but it didn't turn out that way. They were supposed to meet on this island, and have this big battle. And he drew ... pictures ... (over) half a dozen ... but (some) ... were taken to Japan and they were never returned to us ... he was testing out different characters to see what he would like best.[23]

O'Brien's named his giant version of Frankenstein's monster—created by Dr. Frankenstein's *grandson*—the "Ginko."

O'Brien's story concept was formulated in 1958. The following year, George Worthing Yates rewrote the idea, renaming the title "King Kong vs. Prometheus." Again, a boxing match was to ensue in San Francisco. Yates's version was ultimately that which John Beck approached Toho with—the one which transformed into the American *King Kong vs. Godzilla* (also scripted by Yates).

In a sense, Legendary's 2014 *Godzilla* entry represents a culmination of two prior films that never came to pass, each contemplated by American producers and animators during the late 1950s and early 1960s. These never-made movies were titled *The Volcano Monsters*, scripted by Ib Melchior and Ed Watson as of May 7, 1957 (an abandoned precursor idea considered and floated prior to 1959's release *Gigantis, the Fire Monster*), and, of course, O'Brien's *King* "Kong vs. Frankenstein/(Prometheus)"—the film concept for which O'Brien had prepared a story outline and aforementioned line drawings during c. 1958–1962. Here, amazing battles between gigantic kaiju monsters were plotted as staged in San Francisco, which is what ultimately happened in the 2014 Legendary case.

The tale of disappointing events transpiring, and chain of

individuals, leading from O'Brien's "King Kong vs. Frankenstein" idea to Toho's production of *King Kong vs. Godzilla* has been recounted masterfully elsewhere.[24] It will suffice to state here simply that aged O'Brien lost control of his proposal once he had passed it along to those he thought could be trusted to do right by him. The original, resulting Japanese *Kingu Kongu tai Gojira* (produced by Tomoyuki Tanaka, directed by Eiji Tsuburaya) appeared in August 1962 and is quite different in tone from the Americanized version, *King Kong vs. Godzilla*, released on June 3, 1963 (produced by John Beck, new scenes directed by Thomas Montgomery).

How times have changed. Back in 1963 I would have given an arm and a leg to see *King Kong vs. Godzilla* just once at the theater. But fortunately, my dad took my brother and I to see it without need for loss of limbs. Then nearly a decade passed before I ever was able to view it again—televised. But in early 2020, while shopping at a local Meijer, there it was! I spied on a discount rack the *King Kong vs. Godzilla* DVD on sale for only $4.99. So today, anyone can watch it at home at their own leisure! Back in that long ago day, though, all we could glean about the secretive making-of Godzilla came via a handful of black-and-white pulp pages published within *Famous Monsters of Filmland* magazine.[25] Given the wealth of information we have today, a means to watch nearly any classic giant monster film at will—gee, it all seems so convenient now compared to then.

Titans Finally Clash

Each crowned king in America—Kong, as in *King Kong*, and Godzilla, as in 1956's *Godzilla, King of the Monsters*—finally earned their title bout! When the American version was announced in 1963, there was something intangible about this that just seemed so right. However, before comparing in depth the two Kong and Godzilla flicks featured in America with respect to Gaian theory, attention should briefly be devoted to the Godzilla versus Kong films inaccessible to western audiences—the relatively unknown 1962 *Kingu Kongu tai Gojira* as well as Toho's sequels that were under Toho's consideration but were never made.

According to Ishiro Honda, lighthearted enjoyment was the impetus behind the 1962 Japanese original version. There was no intention to portray a "veiled treatise on the state of relations" between the U.S. and Japan.[26] Furthermore, Tsuburaya, who echoed Honda's feelings concerning international relations then, was more focused on "children's sensibilities." Although, according to Arikawa Teisho, special effects assistant, as documented in William Tsutsui's 2004 book *Godzilla on My Mind*, Kong and Godzilla—symbols of the U.S. and Japan, respectively—were intended "to represent conflict between the two countries."[27] There was only a single ending used in both films, where Kong's and Godzilla's climactic battle seemingly ended in a tie. However, as Tsutsui further reflected, America's Kong rescued an "embarrassingly impotent" Japan from Godzilla.[28]

In contrast to the 1963 Americanized version, the Japanese film was a satiric, cynical merging of two gigantic (formerly harrowing) creatures, the first such intentional comedic monster yarn since the slapstick days of Bud Abbott and Lou Costello. Steve Ryfle refers to Toho's monsters as "prehistoric palookas."[29] Yes, Toho's lumbering contestants were now akin to sumo-wresters, instead of heavyweight boxers practicing the sweet science per O'Brien's concept. Critics who have seen both American and Japanese versions seem to prefer the 1962 Japanese film because, well, it simply works better in its original telling. Meanwhile the Americanized 1963 movie was dumbed-down amateurishly in scenes with new footage, while attempting to reestablish seriousness in places where it was never intended. The original 1962 plot, script and musical score are also considered superior.

Toho intended for a grudge match to happen between the two giant monsters as, according to John LeMay, a "repetitive" sequel was planned for Japanese release in 1963. In this unmade movie, audiences would have seen Kong battle a giant scorpion before his inevitable bout with Godzilla. Kong also would have acted as protector for a young child, survivor of a jungle plane crash. Meanwhile, a lifeless Godzilla was to be revived using electricity before the inevitable rumble began! But ultimately both monsters would have been consumed in an erupting Mount Aso.

Three decades later, in drafted scripts, Kong was to battle Godzilla in a film plotted as "Godzilla vs. Mechani-Kong," described by LeMay. There was a distinct, added science fictional twist inspired from the 1966 film *Fantastic Voyage*. An excerpt from LeMay's synopsis best describes where this peculiar idea was going before getting scrapped in 1992. "Godzilla awakens from the ocean floor and then comes ashore in America where it soon becomes apparent that the nuclear reactor within its body will detonate. ... Mechani-Kong injects Godzilla with a miniaturized team of scientists to stop the meltdown ... while the robot ape battles him on the outside."[30] Instead of a volcano, this time the two behemoths perish within the San Andreas Fault.

Of further interest would be the 1975 comic-strip *King Kong vs.*

Godzilla featured in England's *Legend Horror Classics*, and the Bobby Pickett–produced recording of Kong battling a giant sea monster off the Japanese coast, titled *Sounds of Terror*. Don Glut refers to these entries in his *Classic Movie Monsters*.[31] More recently, other talented artists—notably Bob Eggleton, sculptor Joe Laudati and Todd Tennant—have captured the spirit of Godzilla and Kong combat through a number of remarkable artistic renditions.

But then of course we have Kong's own musings, who was then in "retirement" from the movie industry, on the fight that erupted in Tokyo, thanks to Walter Wager's interview with the hairy palooka. Kong stated in *My Side by King Kong: As told to Walter Wager* (1976):

"I gave the silver screen another chance in 1963 when I starred in *King Kong vs. Godzilla*, a harmless Japanese romp that co-starred a young kid with horrendous breath. Fire shot from his mouth, and his posture was atrocious. I taught the kid plenty and gave him great tips such as looking out for short agents."[32]

CHAPTER EIGHT

Colossal Conflict

"Nothing, nobody can stop the great showdown, when King Kong and Godzilla meet to fight for survival of the fittest!"
—from Universal-International's 1963
theatrical trailer for *King Kong vs. Godzilla*

"Never know what you'll run into when you drill holes in old Mother Earth. Sometimes she resents it, and you're apt to find yourself smack in a mess of trouble."
—A prophetic line from *Reptilicus* (Saga Studio/AIP,
1962; directed by Sidney Pink, screenplay
by Ib Melchior and Sid Pink)

While it is nary possible today to find critics or reviewers favoring or, more emphatically, truly loving the Americanized *King Kong vs. Godzilla* (1963), to me, then, it was all good and real. Am I alone?

Nuclear themes had faded by 1962, as "1963 signing of the Limited Nuclear Test Ban Treaty reduced the worldwide threat of nuclear war. As a result, the Showa films largely abandoned the antinuclear message that had driven ... earlier Godzillas."[1] J. D. Lees further notes a later "bright and expansive" nature of the Showa entries, which "often emphasize technological advances of Japan's economic revival."[2]

Cinematic sparring between the world's two greatest kaiju/behemoths—Godzilla and Kong—reflected television antics in contemporary Japan. According to Ishiro Honda—as recounted by J.I. Baker in *Life: Godzilla the King of Monsters*—one stunt broadcast in April 1962 proved particularly unsettling. Audiences tuned into a wrestling match in which "one opponent bit another's forehead, opening a bloody gash."[3] Viewers were evidently appalled, yet en masse simply could not turn away. Baker opines that 1962's *Kingu Kongu tai Gojira*, and its Americanized version, may have been partially inspired by that televised debacle.

Films like 1954's *Gojira* (aside from its prevalent antinuclear message) and *King Kong vs. Godzilla* represent early forerunners of a new class of apocalyptical science fiction, distinctively tapping into *paleontological* sources—another popular muse for authors of modern horror, extending into the new millennium.

James Rollins is one current writer whose novels sometimes delve within the dark theme of man's unwise industry—unearthing forbidden relics out of paleo-history. True to Rollins's canon, when humans penetrate downward into Earth's under-layers and strata, unsettling, sinister events transpire, bordering on the retro-evolutionary. Reciprocally, when primeval biota resurface, humanity suffers. His *The 6th Extinction* (2014) and *The Demon Crown* (2017), for example, present potential for unholy, cataclysmic events stemming from a primeval past unleashed, returning—stoked by man's misguided inventiveness, especially biotech—threatening civilization.[4] Such themes are quite familiar to those reveling in the 2010s Legendary dino-monster films.

In *Demon Crown*, a mystery at the Smithsonian quickly escalates into a crisis. The paleontological scourge turns out to be an enlarged wasp species—originally from the Cretaceous Period—carrying a "Lazarus" microbe, immortalized in amber. Millions of years later, miners in Poland cracked open dinosaur bones preserved in the amber deposit, thus "aerosolizing the cryptobiotic cysts." The microbe and genetic makeup of the wasps allows their kind to regenerate upon exposure to air, after infiltrating and horrifically feeding off a host—such as people—hollowing us out during larval stages.

People can become infected with carnivorous wasp larvae on such a rapidly widening scale that the only solution to end this infestation may be to nuke Hawaii, where the first swarm was unleashed by a diabolical Japanese terrorist group. Can a paleontological breach necessitate a nuclear strike? Consequences posed by Rollins's imagination would seem more dreadful than even if a monstrous, prehistoric towering kaiju were to erupt from a nearby volcano!

Then in his thought provoking *The 6th Extinction*, an ineffable (even to Charles Darwin![5]) "primeval" haunts civilization. Yes—indeed, in Rollins's canon, not unlike *Jurassic Park*'s, life (and not always the "good" kind) finds a way! Or, rather conversely, as Carl Sagan once stated, "Extinction is the rule. Survival is the exception." But as well according to Rollins, terrorists have an edge because "extinction is fast. Evolution is slow."[6]

Here a mad scientist "playing God ... mixing foreign genes into established species,"[7] with motives analogous to eco-terrorist Alan Jonah in 2019's *Godzilla: King of the Monsters*, prepares to unleash organisms (synthetic, genetically-modified yet derived from exotic species, such as hardy extremophiles, or those having survived within recesses of Antarctica's icy tomb for millions of years) to combat and prevail over species currently occupying Earth's mainstream niches, destroying us. Overall, if unmitigated, this would become a red-in-tooth-and-claw, survival-of-the-fittest, law of the jungle, planetary invasion certain to wipe out humanity—comparable in scope to the great Permian extinction! Meanwhile, government officials weigh whether it is necessary to nuke a genetically mutated microbial infestation spreading through California.

Further revealed in *The 6th Extinction* are Rollins's paleo-themed inspirations via homage to classic, paleontological crypto-sci-fi literature such as Arthur Conan Doyle's *The Lost World*, Jules Verne's *A Journey to the Center of the Earth*, with allusions to H.P. Lovecraft's *At the Mountains of Madness*, or John Taine's *The Greatest Adventure* (with a tinge of Crichton's *Andromeda Strain* and a sense of 1933 *King Kong*'s [possibly] "lost" Spider Pit footage thrown in for good measure).[8]

In *The 6th Extinction*, Rollins engages readers with a "Dark Eden" genetic–"XNA" matrix (e.g., exotic and resistant, yet predatory DNA) laboratory ensconced inside a Brazilian *tepui* like that which inspired Arthur Conan Doyle's "Lost World" plateau over a century ago, and an expedition descending into Antarctica's mysterious entombing, encapsulating icy realm (an alternate "ecosystem … stuck in the Carboniferous Period … a fossilized sculpture of an ancient world," known as Hellscape). Rollins was aware of the Wilkes Crater of Antarctica, an asteroid impact event at one time thought to have resulted in the Great Permian mass extinction 250 million years ago. So he further speculated, "If all those environmental niches were emptied out by this extinction, what if that same meteor brought something foreign to fertilize those newly emptied fields?"[9] Hence hellscape?

These settings conceal exotic ecosystems, "shadow biospheres" where flora loaded with unbearable toxins grow and horrific retrobeasts—such as a 12-foot-tall carnivorous, bioengineered *Megatherium* variant—dwell. (In particular, Rollins's tunneling into Antarctica seems conceptually derived from that detailed in his gripping 1999 Vernian novel *Subterranean*.[10]) The Dark Eden organization, advocating global neogenesis, "a natural world beyond humankind, promoting acts of ecoterrorism" is eventually thwarted … *this time*![11] Interestingly, some modern ecologists do indeed speculate, perhaps wistfully, that "a great extinction could lead to new and exciting life-forms, new pathways for evolution, even creating a New Eden." A "reset,"[12] although this clearly would not be a scenario to look forward to in our lifetimes.

Intriguingly, this ever-popular *Jurassic Park*–ish theme gains a new reality check. The strangest thing about science fiction, inclusive of its horror elements, is that it so often evolves into *real* science. And so, as this paragraph was written in March 2020, I was intrigued by two pale-ontological discoveries mimicking what Michael Crichton prophesied three decades ago. First, scientists unveiled the skull of an exquisitely preserved 99-million-year-old hummingbird-sized dinosaur species, found in Burma, encased in amber! Then another publication reported "evidence of proteins, chromosomes and chemical markers of DNA in exceptionally preserved" 75-million-year-old dinosaur cartilage found in a Montana fossil. Therefore, "remnants of once-living (cells), *including their DNA*, may preserve for millions of years."[13] Can a real Jurassic World be that far off in Earth's future? Life always finds a way.

Very dark and somber themes are prevalent in today's apocalyptical fiction—encountered in Stephen King novels, and movies and in tele-vised series broadcast on the SYFY channel, or, for example, AMC's *The Walking Dead*. Unsettling societal changes resulting from the deadly coronavirus pandemic in world history underscores how prophetic and terrifyingly real/*non*-fictional apocalyptical circumstances can truly seem. Or, perhaps, metaphorical, as witnessed throughout Leg-endary's paleo-themed giant dino-monster films of the 2010s, leading toward 2021's climactic *Godzilla vs. Kong*. However, such dino-monster apocalyptical/paleontological metaphor was arguably less apparent, soft-pedaled in Toho's original forerunner *King Kongu tai Gojira* (1962), and the Americanized *King Kong vs. Godzilla* (1963).

Plotwise, 1963's *King Kong vs. Godzilla*, for both monsters a color film production first, rather satisfies, although was merely con-trived to collide Kong with Godzilla in "the most colossal conflict the screen has ever known!" The odzilla in this film, presumably, must be the fire-breathing dino-monster Gigantis resurrected from the 1955–1959 film(s)—dramatically emerging from an iceberg in the Bering Sea. Then Kong arrives on the scene, not on Skull Island, but instead Faro Island. Eventually, after some silly paleo-mumbo jumbo is professed on camera, with both monsters tramping through Japan inexorably toward one another, the two giants climactically clash on Mt. Fuji, in a let-them-fight showdown, with America's Kong (ironically) viewed by some as humanity's defense against Japan's harrowing nuclear-charged menace—Godzilla.

Can Kong stop Godzilla from reaching Tokyo? Which kaiju will prevail?

Onscreen, a well-choreographed pairing of mighty contests ensues: an initial ~2.5-minute display confrontation followed by a prehistoric monster wrestling match for the ages. That indecisive ten-minute outcome still has fans and writers squawking today.

Neglect the appearance of that shoddy giant ape costume: get into the moment!

Round 1

In the Americanized version, as Kong and Godzilla head relentlessly toward one another, Dr. Johnson—the movie's paleontologist—opines that Kong is "sensing presence of a hated enemy." Furthermore, underscoring the prevalent metaphor, "a battle of the giants which may or may not have taken place millions of years ago," is soon to be recreated on Japanese soil. Giant apes versus dino-monsters—has happened before? Sadly, Dr. Johnson can neither confirm nor deny this extraordinary speculation. Faux, paleontological mumbo-jumbo is offered. American reporter Eric Carter startlingly states, "The world is stunned to discover that prehistoric creatures exist in the 20th century." Then Dr. Johnson further clarifies while referring to a children's dinosaur book—one that I still possess: "Godzilla is a dinosaur awakened from suspended animation, a cross between the *Tyrannosaurus rex* and *Stegosaurus.*"[14]

Kong (played by suit actor Shichoi Hirose) moves through a forest, with Godzilla (played by Haruo Nakajima) now visible on a higher rocky ledge, offering a roaring challenge to the approaching ape. Channeling the movie audience's captivation, human observer commentators are already betting on who will win this skirmish—one actor initially preferring Kong's chances.

Kong growls, roars and postures back and forth, pounding his mighty chest. Godzilla gives Kong a sample of what could happen if he isn't careful, blasting an errant helicopter with his fiery ray. Kong, unimpressed, hurls a huge boulder upward at Godzilla. Now a second observer states, "I bet Kong has no chance." A ritual coin toss foreshadows victory for Godzilla.

Kong heaves another boulder toward Godzilla. The latter's massive fins flash: a ray issues forth toward the treeline, singing the startled ape's fur. "Like fighting a flame-thrower." Kong, now backing away defensively, is labeled "chicken" by one of the observers as they plead for him to fight. But now Kong already has had enough and trods away, slightly scorched, his pride wounded. Godzilla mocks his easy win with a triumphant roar, waving his massive arms at the loser of this first round. With Kong in full retreat, scratching his head, Godzilla stalks away. Evidently, Godzilla will not be stopped on his inevitable trek toward Tokyo!

Round 2

Following some human drama and their ineffectual efforts at stopping the Tokyo bound monsters, a soma-induced Kong awakens before a crazy balloon drop over Mt. Fuji. Godzilla awaits. Fierce fighting ensues! Kong challenges, thumping his chest, but then, deciding caution is the better side of valor, stealthily hides behind a rock formation. Godzilla approaches but is surprised by Kong, who yanks his tail. Godzilla, now clinging to rocks for support, shoves Kong away with a quick tail thrust. Kong hurls a boulder and Godzilla responds with a ray burst, once again singing Kong's fur. Alarming, blaring horns sounding from 1954's *Creature from the Black Lagoon* may be heard in the musical score. Suddenly an odd tennis match of boulders batted back and forth by the two combatants commences, devolving into a wrestling, pushing-shoving match ... but neither monster goes down this time. Another ray is fired.

One witness declares, "Can't tell who is winning." Neither can the audience, but at least Kong is battling this time.

The monsters tumble downhill. Kong pounds into Godzilla face with his fists. While this strategy may have worked on a stop-motion animated tyrannosaur three decades earlier in 1933's *King Kong*, Godzilla only rolls away, now pelted by more thrown rocks—Kong pitching like a champ. Kong jumps for joy, then ludicrously somersaults head-first into a rock pile, rendering him unconscious. Godzilla proceeds to bury Kong under more rocks and dirt and thumps downward at Kong with his massive tail. A second or two of stop-motion animation, only, is then inserted showing the suddenly battling pair. More rocky burial ensues after Kong is downed again, while Godzilla roasts the ape (not unlike he finished off Anguirus in 1959's *Gigantis the Fire Monster*). Is it Kurtains for Kong? Nah!

An electrical storm changes the balance of this all-time battle, energizing Kong who emerges from his stupor, zapping Godzilla with electrified fingers. Then Kong stuffs an uprooted tree into Godzilla's mouth. Godzilla spits it out with flame breath, burning branches scorching Kong again. But no longer apprehensive, Kong shakes it off—at the ready for more. Finally, in his moment, heroic Kong can clearly take whatever Godzilla dishes out.

Near the mountain base, more upright shoving combined with

Kong's, er, "electrified" personality causes Godzilla to stumble backward. Buildings are smashed. Kong now clearly appears to be winning on the renewed assault.

Puppet versions of the monsters are spied above the treeline as the tumultuous battle nears Atami Castle. Godzilla trips Kong with a vicious tail swipe, but enraged Kong charges back into the fray as the two crash into another row of buildings. Godzilla fires another ray burst, but Kong, incensed, rushes forward.

The monsters battle their way toward Atami Castle; Godzilla defeated his last prehistoric foe—Anguirus—adjacent to Osaka Castle. Can history, in a sense, repeat itself?

Kong, kicking boulders, drives Godzilla back toward the castle. But there's no safety for Godzilla in defensively situating himself behind this ornate building. With Kong on the rampage—the monsters standing adjacent to one another with the castle in between—it is smashed into smithereens with two sets of huge, powerful paws. The fragile structure crumbles, and while clutching one another, the still-struggling monsters plunge into the sea, in their wake causing a tidal wave and massive earthquake—Mother Earth's veiled response to this colossal combat.

In the distance, at sea, Kong is seen swimming away from Japan; Godzilla disappears without a trace. Meanwhile, Kong is wished luck on his long journey home. So, which giant prehistoric monster actually won?

In spite of rumors perhaps started or fueled by a contemporary issue of *Famous Monsters of Filmland* magazine, persisting for many years, two endings—one for Japanese audiences and another intended for USA release—showcasing favored victors, were *not* filmed.[15] Don Glut debunked this popular "G-myth" in his 1978 *Classic Movie Monsters* book. There's only one battle climax version. Godzilla's disappearance at the movie's end has been interpreted as the amphibious monster's underwater escape into aquatic hibernation, thus he's unseen above the waves. In fairness to audiences' reactions then, it seemed as if Godzilla may have succumbed. Despite all the hype, they fought to a draw!

Kind of a letdown? Afterward, Godzilla's film career catapulted forward, becoming the most famous and popular giant monster of all. Meanwhile, Kong was relegated to fewer performances in ensuing decades. So perhaps Godzilla *was* victorious after all, commercially.

Interlude

It was during the emptying of my mother's house as we prepared for its eventual sale during the summer of 2020 that a most remarkable discovery was made—one that wouldn't have mattered to anyone else, but nonetheless left me overjoyed! I had long blamed my parents for tossing out so many of our old movie monster paraphernalia, many of which would have otherwise become collectibles, during our move from Park Forest to Deerfield, Illinois, in August 1968. I mournfully recall seeing our valuable stuff looking like junk out there on the curb ready for garbage pickup day as we rolled down the driveway and away from that village forever, on to brighter vistas. Yes, tragically, there went all my old Aurora monster models, other plastic monster toys, piles of *Famous Monsters*, comic books, even recordings—or so I thought. The reign of monsterdom had ended for me in Illinois (as seemingly was happening in popular culture around the country while the Vietnam War heated up), or so it seemed.

Eventually, one item magically reappeared upon our arrival in Deerfield as we unpacked; my Aurora Godzilla had survived the journey. Nor apparently had any of those other monster-y things that we had bonded to in childhood smuggled their way into our automobile's trunk on our journey northward. Most of our dinosaur books and toys had been packed and moved, but the monsters had mostly suffered extinction. Nearly age 14, I suppose it was time to grow up partially anyway. Enough with the monster-nerdy things, right?

Then one odd day 52 summers later, as my brother Karl and brother-in-law (Karl's partner) Tanner were visiting the Deerfield house, facilitating removal of things to be discarded via dumpster during that COVID-infested July of 2020, I became startled by several of their discoveries. Upon my arrival I was overjoyed to immediately spy there on a dining room table.

Ah, but first ... *reminiscing* ... for me, fourth grade (which I entered in the fall of 1963) in Park Forest was a lousy experience. That's when one of my most treasured possessions—my handwritten packet of monster stories—was confiscated, *twice*! First by my teacher who claimed I wasn't paying attention in class, and then afterward by my parents who punished me for not paying proper attention to school work. (Guilty as charged!) Thinking back, my grades *were* terrible that year, certainly reflecting my growing addiction to sci-fi and horror monsters. Yes, I was naïve, but to me horror movie host John Zacherle (1918–2016) was a hero, while Forry Ackerman (1916–2008), writer, sci-fi enthusiast and editor of *Famous Monsters*, was a god among men!

Not all of my stories were original, yet in a juvenile way they captured an essence of that long-ago disturbing heyday in Cold War monster movie and televised pop-culture. But then suddenly they were gone—vanished, extinguished. And so for decades I'd forlornly believed the stories were dumped into the trash long ago, never to be seen again. And I moved on. Until July 2020, when Karl found the stapled packet of stories buried in the bottom of a totally stuffed dresser drawer in his old room. Then I found myself reliving the early 1960s heyday of monster-dom once again.

As of 1962, having seen Godzilla and Kong on our black-and-white television screen several times, and reveling in my early fascination for paleontology and dinosaurs, I was already a fan of giant monster movies—my favorite kind. And so naturally several stories I'd written, stapled in my packet, featured Godzilla, Kong and other giant dino-monsters of my own design (e.g., King Colidese). For our purposes here, one such relevant story was titled "King Kong vs. Godzilla (Like it Was in the Movie)."

From my re-reading of this long lost gripping tale, scribbled in no. 2 lead penciled script, it would seem within hours after having seen *King Kong vs. Godzilla* at the theater in June 1963, I fervently dashed out my reminiscences on my little bedroom desk, approximately 800 words

Figure 20: Diorama by the author constructed in 2019 pitting two vintage Aurora toy models in a scene pantomiming Kong's classic battle-to-the-death against Tyrannosaurus in RKO's *King Kong* (1933), with the beautiful Ann Darrow (played by Fay Wray in the film) as the voluptuous "prize." For this piece, the Aurora Allosaurus (at right) stands in for the Tyrannosaurus (model diorama and photograph by the author, 2019).

written on three double-sided sheets of school-lined paper, as much as I could recollect anyway, thus recording (along with six crudely drawn illustrations), my immediate impressions for posterity—an indication of how enraptured I was that day. Years later I was horrified to confront my clumsy, crude efforts to draw Godzilla's fins. (A school friend drew the Kong picture on the last page of the story, a result which I recall being very disappointed in.)

I captured the essence of all the major monster scenes: Godzilla bursting out of the glacier dooming the submarine; Kong's tussle with a giant octopus and invasion of Tokyo; the military's unsuccessful attempt to trap Godzilla in a burning trench in the ground; electrical high-tension wires halting Godzilla temporarily; Kong hoisting a train car; the Japanese carrying Kong (via helicopters and balloons) to fight Godzilla on Mt. Fuji; and that climactic monster fight.

My youthful conclusions of certain scenes may seem odd or amusing today. For instance, Godzilla's "roar sounded like a screech of a million cars." Then when Godzilla pursues a lady (played by Mie Hama) down a hillside following a train wreck (a scene which quite literally gave me recurrent nightmares back then), I wrote, "He was following her for dinner." And upon instance of Godzilla's preliminary engagement with Kong, I stated, "Godzilla seemed to win that fight," which seemed correct, and yet peremptorily (as if my hand was getting tired from all that writing) I scored their final protracted battle in favor of Kong without description or embellishment. "Then King Kong won the fight. ... The End." Nothing was recorded on those back-and-forth hurled boulders or their tumultuous wrestling match.

In retrospect, none of my monster tales were any good; in fact, they were real stink-bombs (sadly even for a 4th-grade level). But writing them was a passion of mine. What's stranger is that now as a senior citizen I can still vividly recall that day, the time we got to see *that* movie at the theater.

True—Kong and Godzilla are for the young at heart. But they're also for the old seeking to become wise.

Godzilla, Kong and, likewise, Gaia are enriched in myth and metaphor. They imbue and exude science fictional idealism, as well. While both Kong and Godzilla are metaphorical, so on many levels is the concept of Gaia—the living planet which our daikaiju are duty-bound to protect, even if humans cannot. In Legendary's finale, Godzilla's apex status becomes defining, while Kong's self-sacrificial eco-preservational actions become broader, planetary in scope.

CHAPTER NINE

Rematch Demanded

Notwithstanding my youthful exuberance with paper and pencil during the spring of 1963 (outlined at end of previous chapter), the first completed novel pitting Kong against a Godzilla-like creature was finally published decades later. Author Alan Colosi, fluent in Japanese, wrote an (as yet) unfilmed 2006 *King Kong vs. Godzilla* movie script, later repackaged and novelized as *KKXG: King Kong vs. Gigantosaurus (First Edition)—The Adventures of Yuriko Kumage During the Greatest War on Earth* (2014/2016).[1] His Gigantosaurus is surely a stand-in for Tokyo's foremost dino-daikaiju. Its morphological description generally correlates with Godzilla, although featuring a "split tail" and sharper, thinner, needle-like spikes (rather than Godzillean, stegosaurian-like osteoderms) along its vertebral column. Meanwhile, particulars of another giant dino-monster in Colosi's colossal novel *Evil Armadillosaurus* would appear consistent with Toho's Anguirus. There is also a nuclear fusion-powered Robo-Kong, quickly dispatched by flesh-and-blood Kong.

In this new origin novel, readers learn of "an evolution of species before all others" on Earth, fashioning a giant dinosaurian (i.e., Gigantosaurus)—although not a "true" dinosaur—that survived (anaerobically) 4 *billion* years ago before being wiped out by an impacting Moon.[2] Fossils belonging to this species, which in that Primeval Era allegedly obtained physiological direct nourishment in electrical energy, were discovered in Arizona and maintained in Area 51's secret military enclosure. However, soon an animate 90-meter-tall specimen emerges from an iceberg in the ocean near Japan, revitalizing itself with energy from lightning strikes.

When attacked, Gigantosaurus issues a white fusion energy ray from its maw. Nuclear weapons therefore cannot be used on this giant saurian-looking fellow because, rather than killing it, the radiation might just be used to refuel its ominous metabolism. "If the creature were to meltdown, it could send a hole through the middle of the Earth."[3] Following Gigantosaurus's skirmish with a U.S. naval vessel, the second titular monster—godlike Kong—is introduced on prehistoric East Island in the Pacific. Here, Kong battles the aforementioned prehistoric Evil Armadillosaurus, a spiky 40-meter-tall quadruped which escapes certain death after burrowing through the ground.[4]

As Colosi states, ensuing battle waged between Kong and the huge spiked reptilian Gigantosaurus represents "probability of the greatest war between two evolutionary legacies. ... But only Captain Yuriko Kumage can make them fight."[5] When the final, titular tumultuous showdown commences on a Japanese mountain slope, at first Kong is reluctant. But sufficient incentive is provided for the hairy god to confront Gigantosaurus in a free-for-all melee that seems reminiscent of the climax—Kong's and Godzilla's *second*, final battle—in Toho's 1962/1963 classic film.[6] Or as Colosi intimates, "it was as if one biodiversity were trying to outdo the other's supernatural rituals."[7] In the end, both "Gods were gone."[8] As in case of the 1963 Americanized Toho movie, Colosi's novel isn't environmentally themed, or entrenched in Gaian concepts.

Then in early 2021, while America, although now armed with viable vaccines, desperately continued to combat the coronavirus pandemic and mutations thereof that had already taken over half a million citizens' lives, finally, Legendary with Warner Brothers released their long-awaited *Godzilla vs. Kong* (directed by Adam Wingard), fortified with another novelization offering considerable background artfully elucidated by Greg Keyes based on Eric Pearson's and Max Borenstein's movie script.[9] This go-around, Kong, who has grown even larger since his discovery during the Vietnam War era, is cast as "not exactly the hero, but at least as the audience identification character—an intelligent being capable of direct communication with humans." Meanwhile, Godzilla reprises his role as an "enigmatic" opposing force of nature.[10]

Adam Wingard, who felt his entire career has led to this pinnacle, offered interesting perspectives on this movie he directed in *Fangoria* no. 359. He believes the "real entry point" into *Godzilla vs. Kong* isn't the 1963 Toho classic, but rather Toho's 1955 oft-derided *Godzilla Raids Again* (Americanized as *Gigantis the Fire Monster*). Why? Because of certain scenes in the 1955 film where Godzilla and Anguirus suddenly are seen to fight in rapid (less ponderous) motion. "Those undercranked effects shots in *Raids* were reportedly an accident during production ... but now they're pillars of the movie's identity."[11] Wingard adds, "the aspect that was most difficult wasn't necessarily the choreography of the monsters. ... It was the initial phases of trying to figure out how to depict scale without slowing down the action ... we had them move just as fast as you would interpret two people fighting. ... Kong punches Godzilla more or less with the speed you'd see in a street fight."[12] In other words, despite wrestling with mechanics of scale and relative sizes, "the solution was to animate the monsters more or less like people."[13]

Other than the two famous giant monsters returning to the big screen together for fighting, there's little interrelationship between the 1963 Toho classic and Legendary–Warner redo. In the new film, the danger-from-extinction-posing radiation theme is even turned on its head, as instead incredible radiation from inside Earth is likened to a life-force that sustained the Titans in earlier geological periods. While the crackpot Hollow Earth theory played no part in Toho's film, plotting in the 2021 movie is entirely structured around this strange yet long-established concept, so historically popular in western (i.e., American and French) science fiction.[14] No effort is made to explain how Earth astronomically developed its hollow, inhabited inner surface.

Thanks to incredible computer special effects, invigorated, vicious battles between giant monsters Kong, Godzilla and Mechagodzilla transpire within Hong Kong's metropolitan area (or at sea on and around battleships), whereas in the 1963 original relatively few Japanese buildings were smashed by costumed combatants via more sluggish-looking suitmation. In the 1963 Toho classic, Kong's and Godzilla's two fights occur mostly in remote areas. But in the 2021 version anything fantastic or thrilling goes, that is, special effects-wise. And radiation, although not from nuclear bombs or emitted from power plants, forms another key aspect of the new film.

An environmental theme *is* evident in the 2019 film (subtle in Toho's original), yet is more subdued in the movie than as outlined in Keyes's 2021 novelization. Accordingly, as in the 2019 blockbuster film, mankind's hubris and overweening tendencies once more are a scourge! But no ... rather than the 2021 film seeming like just a Toho remake, here Legendary's incredible Kaiju/dino-monster series turns full circle, from Godzilla and Kong, in a sense thematically yet reciprocally, harkening back to themes explored in 2013's *Pacific Rim*, with mankind, despite all, seemingly worthy of salvation.

Emma Russell (Vera Fermiga), who died in Boston's great Titan showdown, was essentially correct about roving Titans' ameliorative impacts on Earth's ecosphere, yet woefully wrong about Ghidorah's intentions for them or her ability to control that alpha—Monster Zero. In the intervening three years between near global disaster resulting in Boston's dramatic kaiju showdown and ruination, and man's pending extinction under Ghidorah's wrath, planetary circumstances have generally improved. Godzilla has become a recognized ally of mankind, while Gaia's health seems reversed—on the mend, that is, relative to before the great Titan invasion triggered by Emma's indiscriminate use of the ORCA. As she anticipated, yet also rather ironically, thanks to their measured radiation output, "Deserts were blooming, ecosystems recovering. Titans were hard on human beings, but they were good for the planet. ... And after that, the world had actually gotten better. ... there had been peace and healing." However, in aftermath of near extinction, perhaps predictably given the ways of human nature, "People were forgetting, returning to the practices that had screwed up the global ecosystem so badly in the first place."[15]

Meanwhile, Godzilla—now affirmed "steward of the global ecosystem"—has been on patrol, a command-and-control phase referred to as the Big Cleanup, where he checked on and corralled the other remaining Titans, urging several back into suspended animation, following some battles (as recorded in a 2021 graphic novel and in Keyes's novelization). Such conflict among Titans was necessary because, as Keyes explains, "a few of the titans had clearly had their fingers crossed behind their backs while they were bowing,"[16] that is, at the conclusion of 2019's *Godzilla: King of the Monsters*, following Ghidorah's defeat in Boston. Behemoth, Scylla and Amhuluk needed to be put in their place, for example—battles not witnessed in Legendary's movies.

The fact that Kong was sealed by then in an observational, protective biodome, proved satisfactory to alpha–King Godzilla—no further encouragement or intervention needed in the vicinity of Skull Island. Meanwhile we learn that the ecosystem on Kong's Skull Island had been ravaged in the wake of intense flooding and storms conjured up in the wake of Ghidorah's assault on Earth. A young deaf girl, Jia (Kaylee Hottle), who befriends Kong is the sole surviving member of the ancient Iwi tribe. Skull Island is known to be an anomalous site at which Hollow Earth had sprung, or flowered upward toward the planet's surface. Meanwhile, Kong has already attempted once to breach the biodome's shell. It will happen again, perhaps successfully next time, unless scientists find a better, more satisfactory home for him.

Anthropologist Ilene Andrews (Rebecca Hall) claims (and later proves) there's an ancient rivalry between the alpha Titan clans represented in modernity by Kong and Godzilla. This is why they seem to be hyperaware of each other's movements—stemming from a war fought between the two clans, when mankind was in its infancy, impressing itself upon the psyche of the ancients, possibly passed along through generations via "genetic memory." Gifted with such an uncanny ability, and given that Kong knows sign language, allowing him to communicate intimately with humans, this godlike Titan's intelligence is at least on par with ours.

Is this not then a most convincing movie-scripted representation of Carl Sagan's subconscious "nighttime stirring of the dream dragons ... the hundred-million-year-old warfare between the reptiles and the mammals"[17] ... metaphor writ large for the cinema? For on the verge of their first combat in the Tasman Sea, "Kong had sensed him before, many times. Sometimes it had been like an itch, but deep inside. ... He had never seen him until now yet there were no surprises when he did. ... The shape of the enemy was like nothing he had ever seen, much less fought; but just the scent of the creature made him angry, and everything fit into a hollow spot inside of him, as if something had been taken out long ago and left empty until now."[18] From Godzilla's perspective, his "territory was invaded. Not by the hidden one, but by the Other, another ancient enemy, older than the three-headed one, a rivalry written into his very blood and bone."[19]

Establishing inherited genetic memory in the story as a real biological thing beyond the science fictional becomes a means for suspending disbelief both in the film and as further elaborated in Keyes's 2021 novelization. As Nathan Lind (Alexander Skarsgard), branded as a pseudo-scientist for his wild claims about Hollow Earth, claims,

> Genetic memory. ... I think he (i.e., Kong) has a map of this place (i.e., Hollow Earth) built into him, whether he knows it or not. Look—when loggerhead sea turtles are born on a beach in Florida, they take an eight-*thousand*-mile trip around the Atlantic basin. With no one to guide them. ... The mothers lay the eggs and leave. But these little turtles know where to go. And they steer using variations in the Earth's magnetic field, until they—the females, anyway—end up back up on the same beach where they hatched to lay *their* eggs. This isn't learned behavior. It's hardwired.[20]

Likewise, presumably, would be Kong's presumed ability to chart his way back through Earth's interior to his ultimate place of origin—and the place where a strange energy beacon may be appropriated by Apex Cybernetics for diabolical purposes.

Meanwhile, Kong evidently trusts Jia more than anyone else, not necessarily all mankind in general. It's as if Kong has become an ecological steward for one particular, now (thanks to Emma Russell stirring Ghidorah's stormy ravages) near-extinct facet of the human race—Jia, sole surviving member of the ancient Iwi. Whereas Kong has remained near the opening of the Skull Island Vortex all his life, sea-faring Godzilla has traveled over (and even tunneled through) the planet. How will Kong's persona and rather delimited view on the state of our world mesh with Godzilla's more far-ranging more cosmopolitan perspectives when they eventually meet—thus reigniting their primeval blood-feud?

Round 1—"Welcome Aboard!"

"One will fall," according to the movie's tagline. But who? An amazing, initial eight-minute battle ensues.

As anthropologist Andrews states, recalling *Godzilla: King of the Monsters'* conclusion, "Kong bows to no one!" However, Kong's and Godzilla's initial Legendary–Warner Brothers skirmish doesn't go smoothly for the simian Titan. For one, Kong is disadvantaged because

Figure 21: Small Kong toy licensed through Legendary's Monsterverse by PlayMatesToys.com, available in 2020, representing an aged, battle-scarred Kong appearing in *Godzilla vs. Kong* (author's collection).

he is half-drugged while on ocean cargo transport (with a full naval escort for his protection that proves futile). Further, Kong, who thinks he may be going to a (new) home, is not adapted to submerged amphibious fighting tactics; Godzilla's forte.

Godzilla swims along the sea surface toward the fleet, despite the barrage of rocket fire attempting to slow his relentless advance. Jagged fins plow through and bisect one naval destroyer on his approach. Rising from the water, Godzilla then capsizes the cargo ship to which Kong is enchained. Kong is drowning! But after a heart-pounding moment, indecisive Lind sets him free, remotely unlocking his shackles. Kong climbs out of the surge, pounds his mighty chest and leaps toward the deck of an adjacent aircraft carrier. Godzilla mounts this ship as well and with both monsters standing aright, quickly receives Kong's roundhouse right to the jaw, causing the huge saurian to reel backward from the blow. Godzilla counterpunches, a more powerful blow sending Kong sprawling on the deck. Then another Kong punch and heave shoves Godzilla back into the drink. Are we witnessing Ali vs. Frazier, or Rocky Balboa vs. Apollo Creed?

But Godzilla's vertically-directed incandescent ray from beneath the waves melts upward through naval steel like butter; Kong bounds into the sea, avoiding being burned to a crisp. Kicking his way through depth charges, used to distract Godzilla, Kong reaches the cargo ship ... where gasping for air he collapses. With all systems shut down—a ruse—Godzilla is satisfied for now, beguiled into believing he has defeated Kong, and remains Earth's sole Alpha.

Or is there yet another contender for the crown?

Unto the Breach

Godzilla's cerulean ray is likened to and derivative of radiation discovered emanating from Earth's core into the Titans' former Edenic inner world—Hollow Earth. But this form of radiation is also generally new to science. "Godzilla *converted* conventionally understood radiation into some other form of energy which manifested into the beam of unknown energy he discharged from his mouth." This is "another sort of energy, perhaps not nuclear in origin, but tied more closely to quantum states. ... Which did not closely resemble the nuclear particles and waves discharged by a fission or fusion reaction."[21] So Apex Industries Cybernetics magistrate Walter Simmons (Demian Bichir) and associate biotech genius Ren Serizawa (Shun Oguri), son of Dr. Ishiro Serizawa who appeared prominently in Legendary's 2014 and

2019 movies, must absolutely tap this elixir in order to vanquish Godzilla, thus capably fueling their own artificial giant monster—Mechagodzilla—with powerful energy ... energy representing the very "wellspring of life."[22]

But penetrating into Hollow Earth's surfaces through the "Vile Vortex" (distorted in space-time) is difficult almost beyond comprehension, and recalls how invading Kaiju breached a barrier portal from deep within our planet to enter our oceanic realm and surficial world from an Anteverse in *Pacific Rim*. However, in *Pacific Rim*, to us, the Anteverse is a hellish landscape, whereas in *Godzilla vs. Kong* breaching the Vortex leads to an inner paradisiacal, primeval world. Although all signs indicate Hollow Earth's environment may be habitable, in order to reach it, explorers must find some means of surviving a crushing gravitational inversion, "A whole planet's worth of gravity reversed in a split second ... like flying into a black hole."[23] Acceleration rapidly rises dangerously—having killing pilots before. While passing through this anomalous membrane, time itself seems to slow. The only way to pierce the veil is via Apex Industries' newly designed Hollow Earth anti-gravity vehicles (HEAVs), which perform admirably.

One interesting yet mysterious tie to *Godzilla: King of the Monsters* is the relative occurrence of Ghidorah's ORCA exhumation site, in proximity of the Antarctic tunnel at Monarch's Hollow Earth Launch Station leading Kong accompanied by Nathan Lind, Ilene Andrews, Jia, and Maia Simmons (Eiza Gonzalez) piloting HEAVs into the Vortex (much larger and more accessible than the Vortex underlying Skull Island). Ghidorah's location when freed from the ice by explosives is in fact nearly adjacent to where Monarch constructed its portal to the interior domain, which is perhaps more than just an interesting coincidence. "Was Ghidorah going into the Vortex, or had it just come out of it? Or neither? The ice all around ... is more than thirty million years old. The ice around Ghidorah was younger ... it clearly melted quickly and re-froze quickly."[24] One wonders then, in Legendary's kaiju canon, did "alien" Ghidorah actually descend from the stars, from above, or instead through some sort of interdimensional "portal" deep within Earth? And when, exactly?

Mother Earth's Womb

As anticipated, Kong, with incentive provided by Jia that there "could be more like him inside," guides Lind's HEAV fleet safely through the frigid Antarctic Vortex into Earth's inner-annular interior. Kong's appearances in a long-deserted Hollow Earth temple—one his ancestors once ruled within—are sequenced at a fast-moving pace. There Kong is intrigued and maybe overjoyed by the lush forests and waterfalls, the odd lighting, vast mountainous subterrain and weird gravitational effects—objects floating in an inner-world sky. (Like a weird, psychedelic *Yes* rock album cover.) He combats a large bat-winged, serpentine-tailed creature (described by Keyes in his novelization as Camazotz) and plods on past a large scaly quadrupedal reptiloid (which gobbles up a strange crab-like organism). Is this where life began? Well, if not, at least from Kong's explorations of a past civilization and the aforementioned temple, this is where the Kong-species originated, and perhaps other giant monsters as well.

Kong spies a Kong-like statue then discovers a bone-club axe exhumed from the skeleton of a fossil Godzillean creature; clearly Kong is destined King of this underworld! Maia Simmons notices the axe is drawing radiation from the core, and so taps this life-force radiation emanating from below the ancient throne room vault. This is the fuel that will soon power her father's monster. As batteries charge, the strange energy is transmitted upward through the Vortex to Apex Cybernetics systems in Hong Kong where Ren Serizawa seizes controls of a fateful monster—a prospective Alpha thus far tested only with moderate success against Skull Crawlers bred in captivity.

And thus, Round Two commences.

Round 2—Don't "Axe" Me Who Wins

Meanwhile, Godzilla remains on the prowl for Apex Cybernetics' fabricated Alpha monster contender. Fast-paced action further leads to the next dramatic showdown, another extended and brutal, bloody battle. Tool-using Kong emerges, wielding his bone-axe club weapon, from the thousand-mile-deep molten shaft leading into Hollow Earth, below Hong Kong, that Godzilla has incredibly created with his ray. Kong summons Godzilla's attention with a vicious growl, then rushes in. Godzilla dodges and Kong crashes into a high-rise structure. Godzilla quickly gains advantage biting into Kong's hairy neck, but Kong pulls free. They wheel around. Kong slugs Godzilla's head, landing a haymaker. He kicks and pushes the saurian's neck and head, averting a fiery ray blast. Kong expertly feints left and right, averting Godzilla's fiery maw. Then Kong stuffs the bone end of his club axe down Godzilla's maw to stifle another burst. When a ray is blasted outward, Kong shields himself from it with his axe-club blade, which is actually part of an ancient, fossil, Godzillean fin—capable of withstanding and absorbing the tremendous heat and energy.

Kong is shoved into the bay as Godzilla sends the club-axe flying toward him like a giant boomerang ... that misses, instead wedging itself in the side of a building. Kong is now retreating, bounding from building to building, Godzilla's heat ray insidiously following his every move, just barely missing.

Kong rips a saucer-shaped structure from a building, holding this frisbee like a shield to reflect the ray, then reclaims his bone axe. But this time a blast from Godzilla's maw proves overwhelming. Kong is slammed rearward into a building while Godzilla is rocked backward on his heels. Watching omnisciently from above in a HEAV, Lind proclaims, "Looks like Round 2 goes to Kong." A premature decision?

Round 3—Don't Tread on Me

This swifter round resulted in what ordinarily would have been Kong's demise, if it were not for human intervention. Kong still looking strong at first atop a skyscraper hurls a projectile. He then leaps on Godzilla, homage to the 1933 RKO classic—that great stop-motion animated tyrannosaur battle. But the Super-saurian flings Kong hundreds of feet down a street, damaging the ape's left shoulder. Godzilla stomps his foot triumphantly on Kong's mighty chest in victory. Despite his anger, growling, wounded Kong cannot answer this final challenge. This time he can't get up, but will not bow. Unstoppable, invincible Godzilla thuds away leaving Kong to die alone. One must fall. Round 3 and match apparently goes to Godzilla.

But while stirrings are continually in motion under the nearby Apex facility—that we'll come to shortly—Jia can feel the weakening vibrations of Kong's heartbeats pounding along pavement through her feet. This leads to Lind's and Andrews's attempt to jump-start Kong's heart using electrical shocks from batteries in their grounded HEAV—a device ironically created by Apex Industries, now utilized for a Titan-saving purpose. The strategy works and Kong miraculously revives—a sign of Providence, perhaps—just in time for a final battle—one that will decide Earth's ultimate fate!

But meanwhile, underground, relying on a sample of Hollow Earth's vital radiation, Ren Serizawa has been charging up his mecha-monster construct—Mechagodzilla, a chimeric Ghidorah-human melded creature that will allow Walter Simmons to become mankind's dominant alpha aberration, even more powerful, yet far less scrupulous than Godzilla.

Ren Serizawa intends to psionically control Mechagodzilla in combating Godzilla—the monster his father had ironically revived in order to save the world. Ren had wielded Mechagodzilla rather successfully in test trials against Skull Crawlers. And he had harnessed the "telepathic potential" of two Ghidorah skulls as major components in the control system. However, Apex had moved the project forward "without ever *really* understanding how and why ... (it) worked."[25] Yes, Serizawa's comprehension of Mechagodzilla's limitations does recall Ian Malcolm's concerns about the mad use of misunderstood genetic power in the original *Jurassic Park*, which led to the thoroughly breached *Jurassic World* theme park. But unlike the more cautious Malcolm, Serizawa, thirsting for power that only Ghidorah's skulls may exude and instill in his brain, throws caution to the wind and barrels ahead anyway with a Mechagodzilla newly amped up with surging, potent Hollow Earth life-nurturing radiation.

Donning the psionic helmet, suddenly Serizawa viscerally feels the power in his Gojira—the one that will destroy the natural version. And yet there was more, "something was also entering him, oscillating, a feedback loop between his own consciousness and the AI (i.e., artificial intelligence). He felt a million years of rage rising in him, hatred that transcended time and space. He felt as if he was sinking into it, dissolving, as another mind full of terrible, alien thoughts began to take his place. ... as Ren died, its vision sharpened. It felt its hands, its legs, its fins, everything. And it saw a shape, a tiny shape, staring at it. ... One that believed they controlled."[26] Victimized Ren Serizawa has allowed a critical mental breach to occur by linking his brain to the Ghidorah–Mechagodzilla mind-melded system charged by an unknown form of radiation beyond scientific comprehension.

Inadvertently, Apex Cybernetics has conjured the ancient antagonistic spirit of Ghidorah, although fecklessly seeking to control it! Now in the embodiment of Mechagodzilla, Serizawa—beyond expired Water Simmons's dark aspirations—has become the Alpha! Simmons refers to this outcome, as he believes it will play out, as Providence. Really? One doesn't deterministically manufacture or easily control one's own destiny or providence. It's all about how things shape out, perhaps randomly, chaotically, or in some sense seemingly "divine."[27]

Round 4—Share the Throne. An End to a War?

Startling the human crowd, Mechagodzilla—the now hulking living supercomputer—erupts from the demolished Apex underground lair headquarters to confront Godzilla, who is disadvantaged after just having defeated Kong in a spectacular title bout. Both monsters have energy rays, but—surprise!—Mechagodzilla's is more powerful.[28] From the get-go, Mechagodzilla is on the verge of winning, flinging Godzilla around like a ragdoll.[29] Mechagodzilla punches Godzilla, driving him backward, firing chest rockets. In fact, Godzilla, evidently on his last legs, is being stomped upon by the giant robo-monster. There is no Mothra to save the day this time. However, Kong, representing a last hope for dominance of natural alphas and humanity, is restored. When Mechagodzilla drags and hurls Godzilla into a pile of rubble, and is about to deliver the final blow with his superior energy ray, an enraged Kong rushes in, joining the fray!

Meanwhile, in the damaged Apex control room, young Josh (Julian Dennison) frantically tries a last-ditch, hail-Mary ploy to slow Mechagodzilla's apparent path to ultimate victory. Josh snatches a hip flask of booze from conspiracy theorist Bernie Hayes (Brian Tyree Henry) just before he hoists it to his lips, signaling mankind's seemingly inevitable defeat, and desperately casts its liquid contents over the panel and keyboard, like a priest sprinkling holy water over a demonized soul. Suddenly, outside in the arena, Mechagodzilla's red eyes flicker; momentarily, the mecha-monster shorted out. Not for long, a mere instant, but allowing sufficient time for Godzilla to regroup and for the two formerly warring Titans to gain a slight edge. *This* is Providence!

Godzilla and Kong send Mechagodzilla sprawling. Kong retrieves his club-axe, clubbing Mechagodzilla, severing a robotic arm. As Mechagodzilla ineffectively tries to reestablish itself from the misfire, Kong rips off its head triumphantly, roaring in victory. Now realizing what they've accomplished together, the victorious Godzilla and Kong tag-team, two former opposing combatants, share mutual respect for each other. The throne must be shared. Godzilla is seen swimming offshore, reciprocally as witnessed in the 1963 Americanized classic (where only Kong was shown swimming away). Are the two alpha daikaiju now sworn to prevent uttermost folly of mankind through forthcoming events? Only future exploits would let us know.

In defeating a newly styled Mechagodzilla (rather akin to those robotic Jaegers controlled by Drift and "neural handshakes")—Man's faux-kaiju creation—Titans Kong and Godzilla, Earth's protectors yet instinctive adversaries, assert their dominance. *They*, not Man, prove to be *the* Apex species after all. And by teaming up for the decisive victory, they've given us another chance in spite of ourselves and all that we've done (or will do). Furthermore, if man hadn't *twice* interceded—first by Nathan Lind and Ilene Andrews representing team Kong, and later by young Josh, Bernie and Maddie of team Godzilla—Godzilla would have killed Kong, and Mechagodzilla would have destroyed Godzilla (and likely a revived Kong if battling solo as well). And with Earth's two natural alphas dispatched, our means of preserving a life-sustaining biosphere would have forever remained in jeopardy—representing an end to Gaian (negative) feedback. Intricate teamwork and cooperation were necessary in order to spare mankind.

And so that's what this new blockbuster film is all about. It's never been my practice to rate movies, but I can't help wondering (as my long-ago, starry-eyed self once did) ... gee whiz (for its impressive special effects) ... do you think *Godzilla vs. Kong* might win an Oscar?

CHAPTER TEN

Non-Classical Daikaiju

Godzilla in New Millennial Alternate Apocalyptic Media

"What makes us human is we can choose to lose the battle."
—(Haruo in Toho's *Godzilla: The Planet Eater*, 2018)

As Bill Bussone stated in *G-Fan* no. 127 (p. 43), "an apocalypse isn't so traumatic for those who come after it as it was for those who went through it."[1] But then *when* exactly is the apocalypse essentially "over"?

Besides Legendary's string of fantastic movies, the new millennial, apocalyptical-themed Godzilla mythos was also recast in graphic novels, as well as three animated films produced by Toho in this period. These original artistic perspectives featured a Godzilla never witnessed before as, say, during the Showa series over half a century ago. While this additional pair of graphic novels may be viewed as further set-ups for Legendary's eventual "ape versus dino-monster" apocalyptic battle, the three Toho films do not explore or lead toward that thematic terrain. While the lesser-known novels trek toward an eventual (filmic, Legendary's) apocalypse, the trilogy of Toho anime films reveal a remote aftermath to our pending cataclysm.

The first Legendary–Warner Brothers official graphic novel publication was titled *Godzilla: Awakening* (2014), written by Max Borenstein and Greg Borenstein, illustrated by Eric Battle, Yvel Guichet, and Lee Loughridge, with cover art by Arthur Adams.[2] Here in this prequel to the 2014 movie we learn more of the elder Serizawa's (i.e., father of the Dr. Serizawa character in the 2014 Legendary film) early post–World War II confrontations with the titular monster, and also further background on MUTO foes. It is in the shambles of an irradiated Hiroshima on August 6, 1945, that he first sees "a monster summoned by our own (i.e., man's) monstrosity." This thing—a gigantic flying bat-like mutation that raids ships at sea—is first seen by the elder Serizawa in August 1946 on a Pacific island. But strangely, Godzilla—referred to as "Gojira"—is already in pursuit of this monster, trailing it along its several appearances throughout the late 1940s and early 1950s, and then strangely warding it away from human occupations.

Finally, in 1953, Serizawa becomes more fully indoctrinated into the nature of the biological problematica posed by these destructive and monstrous cryptid creatures that occasionally appear. As a representative of the Monarch Cargo Company, Serizawa obtains more background on this flying bat-like creature, known as the Parasite Shinomura, which has menaced a number of sites and facilities since World War II. He encounters evidence that this giant creature—a super-organism—grows from a single giant cell into a "swarm of death."

Realizing Shinomura was thwarted by its ancient nemesis Gojira on several occasions, Serizawa reasons that the dino-monster's origins date much farther back than recorded antiquity. Theirs is an "eternal struggle. A balance." In fact, Serizawa claims the great era-ending Permian asteroid impact (for which there's little actual geological evidence) causing mass extinctions 251 million years ago on an unequaled scale, somehow forever "diminished atmospheric radiation," forcing such monsters through subterranean caverns—closer to radioactive sources nearer Earth's Core. Geological eras passed by until mid–1945 when mankind weaponized the atom, detonating atomic bombs, inadvertently summoning a Shinomura survivor toward Earth's surface. "We lured it up with radiation," Serizawa posits, "it came looking for more ... a world littered with radioactive material, a new habitat created by us. ... We brought this on ourselves." Feeding on radiation, it continues to grow.

Serizawa surmises the only thing that can defeat it is Gojira who has an instinctive, genetic loathing of Shinomura. The two monsters are ready to wage their timeless war at the Bikini Atoll in March 1954— when the United States' Castle Bravo tests detonated a hydrogen bomb resulting in the historical *Lucky Dragon* incident. And yet the Parasite seems to have been obliterated in the blast.

Next, in *Godzilla: Aftershock* (Legendary Comics, *The Official Prequel Graphic Novel to Godzilla: King of the Monsters*, 2019),[3] written by Arvid Nelson, illustrated by Drew Edward Johnson, we learn more of Dr. Emma Russell's background before she turned eco-terrorist. Also, we are introduced to Alan Jonah, who believes "the monster is man," and staunch adversary to Monarch, with whom Emma eventually collaborates.

This short graphic novel is thought-provoking, promising a block-buster of epic proportions:

> The King of Monsters becomes an endangered species as an ancient terror rises from the depths of the earth, unleashing a series of devastating earthquakes, and driven by an unstoppable primal instinct that will test Godzilla like never before. Monarch Operative Dr. Emma Russell races to stop the threat, as clues emerge to reveal a terrible secret—a tragedy of apocalyptic proportions from the distant past that changed the course of human history is returning to threaten it once more. A shadowy figure (i.e., Jonah) stalks Emma's every move as she travels the globe to uncover secrets, while Godzilla clashes in an ancient rivalry as old as the Earth itself. The fate of the world lies in the balance.[4]

Set in 2014 during a period following the MUTOs' 2014 attack on San Francisco, against Godzilla, Emma has come to realize that Godzilla has a natural protective instinct; he is a "guardian." And by now, men are realizing that formerly divisive matters such as religion, overpopulation, and territorial disputes seem relatively insignificant because "humans are at our best when we're facing a common threat." That "threat" looming as apocalyptic monsters manifests out of antiquity. Such monsters are being discovered, disinterred and correlated to immense boreholes of distinct geological ages found all over the globe!

Godzilla's foe is another massive MUTO-like bug covered in a roach-like carapace. According to Phoenician relics recorded on ancient stone tablets, Godzilla—possibly a fallen "Gojira-type ancestor"—was referred to as "Raijin," or the "Divine Beast" or Great Dagon (a Lovecraftian[5] name) during the 11th century BCE, when it fought the MUTO-like bug to a near certain death. This variant MUTO was known as "Jinshin-Mushi" the Dragon-Beetle, a creature for which divine temples were erected in some localities. Like in the 1979 20th Century–Fox movie *Alien* (directed by Ridley Scott), the female bug injects eggs through ovipositors into Godzillean bodies—deposited into one of those huge caverns/chambers she has dug, where its young are incubated. The larvae or hatchlings feed off radioactive energy harvested within the Godzilla prey. Over time the bug-like parasites eventually metamorphose/evolve into the next form in a chain of "MUTO Prime" subspecies (rather like how microbes mutate as they propagate through generations of infected hosts).

According to Emma, comprehending biology of these ancient creatures through some mental gymnastics, the "Jinshin is the parent super-species to the MUTOs" or the kind that attacked San Francisco. The MUTO is a genetically perfected parasite: "Evolution has molded it for one purpose, to implant its eggs in Godzilla to *kill* him." The eggs feed off of uranium-enriched hemoglobin in Godzillean blood, thus draining the Titan of its vital radioactive sustenance, resulting in its death. Then the embryos grow over centuries before emerging from the host.

Eventually, when MUTO subspecies are spawned, in consequence to parasitizing the Godzillean lineage through geological time, their emergence from those ancient boreholes results in massive upheavals on Earth. In fact, it seems the ages of those boreholes, burial chambers, do correlate with timing of geological mass extinctions and crisis episodes in human history. So, Emma forecasts what unfolds in the 2019 Legendary film. "If we can't stop the current MUTO Prime, we're looking at environmental and tectonic upheaval, potentially on a massive scale." More to the point, "If we don't help Godzilla, we are looking at a global extinction event that ... will make San Francisco's 2014 war zone look like a minor traffic accident."

Generating eardrum-splitting sound reverberations from a crudely improvised musical device (i.e., mimicking sounds of pulsing bug eggs implanted within a Godzillean host), Emma attracts the MUTO, which disastrously goes on a rampage. Her biosonar technique evidently requires further fine-tuning before it can be used effectively. The novel's climactic battle between MUTO Prime and Godzilla takes place in Montana, at a decommissioned nuclear storage site, a source of radiation which attracts Godzilla. And yet Emma's jerry-rigged device facilitates Godzilla's ultimate victory.

Next, Toho, in conjunction with Polygon Pictures and Netflix, released three computer-animated Godzilla films between November 2017 and November 2018. These were also of an apocalyptical outcome, although plotted in Earth's distant future, 20,000 years from now. The first of these, *Godzilla: Planet of the Monsters* (directed by Kobun Shizuno and Hiroyuki Seshita; screenplay by Gen Urobuchi), introduced two friendly star-flung species—the Exif and Bilusaludo, religious missionary cultists and logical, tech-savvy (Klingon-like) aliens, respectively—who attempt, unsuccessfully, to rescue humanity during the closing of the 20th century from a nightmarish giant kaiju invasion.

While monsters emerge from the "depths of Earth, sea and forgotten legends," it is the added arrival of Godzilla—"avatar of destruction"—that causes Earth to collapse. Unimaginably, Godzilla even survives one salvo of 150 thermonuclear bombs! When the aliens' facilitation to destroy Godzilla fails, the rest of humanity joins them on a trek aboard spaceship *Aratrum* to find another habitable planet. In this renewed universe setting, however, as one of the Exif, named Metphies, divulges, "creatures like Godzilla" viewed as wrathful, divine, "vengeful hammers for arrogance" are to be found throughout the universe. The planet of Exif origin was destroyed by such a monster.

Eleven light-years outward, "drifting for 20 years" through space, searching for a second planet where they could safely land and reestablish civilization still hasn't been found. The odds of finding such a destination seem narrow and, meanwhile, their imprisoned Captain Haruo Sakaki publishes a manifesto on how Godzilla could be destroyed—should they decide to return to Earth. That decision is eventually made and, via warp drive, they return to our Solar System. However, due to the physics of space-time general relativity, 19,200 years (based on a carbon-14 measurement) have elapsed on Earth since their late-20th-century departure! Seemingly immortal, Godzilla still lives!

And Earth's ecology has transformed, evolving considerably, catalyzed by Godzilla's presence. Biota living on the surface have become fully adapted to Godzilla's "phenotype." This is quite evident when the *Aratrum*'s ground reconnaissance team suffers attack from a flock of gigantic metal-clad, bat-winged dragons (called "Servums," resembling Charlton comic's Reptisaurus of the 1960s, yet possibly stand-ins for Rodan). However, despite what has occurred, apocalyptically, during the intervening twenty millennia, Haruo notes that this planet, however transformed, "will always be Mother Earth." This remark is uttered despite the perspective that Earth *chose* Godzilla as its "lord of creation" instead of humanity. Haruo reflects that global warming, detonation of nuclear weapons, emissions of carbon dioxide, and repeated environmental damage *fostered* the arrival of monsters.

Haruo has analyzed Godzilla's fighting characteristics, believing he may have found a weak point, an anomalous genetic vulnerability that may make it possible for them to destroy the monster. That vulnerable organ happens to be a dorsal fin that generates an asymmetric electro-magnetic pulse shield protecting Godzilla. Haruo's vendetta against Godzilla stems from how the monster has stolen man's "justice, faith and basic human dignity and pride." Now he wants to finish "immortal" monster Godzilla once and for all and then reclaim Earth.

Haruo's device and strategy ultimately succeeds, but it turns out though that this defeated odzilla is but an offspring. Then the original Godzilla, now much larger at 1,000 feet tall—six times taller than originally encountered 20,000 years ago—emerges from the rubble! We later learn that Godzilla's heat ray is so powerful that it could potentially penetrate the atmosphere from the ground, and destroy the orbiting *Aratrum*.

In the sequel anime film, 2018's *Godzilla: City on the Edge of Battle*, a Mechagodzillean remnant—the titular city laboratory in which that monster was developed millennia before—appears in the fight against the original Godzilla. We learn two interesting details about how Earth has progressed since its abandonment by the last humans. First, the landing team discovers a tribe of humanoids known as the Houtua, surviving mostly underground yet adapted to the toxic atmospheric surficial conditions as well. Interestingly, the Houtua are instilled with an insectoid genetic marker. These people belong to a "cult of the egg"— alluding to Mothra (who otherwise doesn't make an appearance here). There are even a semi-telepathic pair of twins, perhaps long descendants of the Infant Island fairies. Secondly, liquid nano-metal (a living form of enzymatic micro-metal), of which Mechagodzilla was fabricated, has proliferated and evolved in 20,000 years—threatening to consume Earth.

This sequel strays into more philosophical, sometimes rather convoluted terrain than the first entry, with the Bilusaludo joyously becoming as-one, fused with the nano-metalloid City. They believe that in order to exert their control over the environment that they must "attain the goal of becoming Godzilla." And as it is claimed, "monsters are called that because they cannot be defeated by man." Ultimately, though, their goal remains to destroy Godzilla by injecting an EMP harpoon into its key dorsal fin, thus allowing humans to re-modify the environment and ecology, making it more suitable for habitation. The familiar Mechagodzilla—referred to as the Bilusaludo's Earth-invasion weapon—never appears, however. Instead, three derivative, human-operated flying "mechas" named Vultures made of nano-metal stand-in for an active Mechagodzilla. Liquid nano-metal floods a capture point for Godzilla, hardening, but Godzilla bursts from its encasement, melting Mechagodzilla City.

Next, with the Bilusaludo's nano-metal Mechagodzillean attack thwarted, in *Godzilla: The Planet Eater* (2018), the Exif's diabolical, deceitful plan to destroy Earth is orchestrated. And their weapon is a mighty god, a planet-destroyer from an alternate universe ruled by different physical laws—Ghidorah, dramatically summoned and controlled by Metphies! Why would Metphies invoke such a suicidal, cataclysmic end to our planet, given that he'd befriended hopeful Haruo? There's quite a bit of philosophical babble interspersed throughout this third, final entry, but his (twisted) reasoning unfolds thusly.

To Metphies, the alien missionary whose own world had been destroyed by Ghidorah long ago, destruction is a blessing, representing a dark "day of harvest," a natural cycle repeating itself on other planets. Furthermore, all creatures in the universe are intended to be sacrificial offerings to (their) *god*—Ghidorah! He claims, "There is no such thing as eternity; everything is finite—find peace and comfort with destruction of the universe." Furthermore, the human civilization that "gave birth to Godzilla stepped over their threshold and so were destined to be destroyed" anyway.

In an earlier scene, one of the human characters muses in a rather Gaian/Medean context whether "if everything we created, all the technology, was not for humanity, but to give birth to those creatures we call 'monsters.'" Are humans really only puppets destined to "give birth" to giant metaphorical monsters threatening apocalypse? Godzilla is mere punishment (of our own design) in retaliation for our arrogance, yet Ghidorah represents divine victory, the "golden wings of a beautiful demise." Then is this why supreme Ghidorah—the planet eater—would seek to destroy Godzilla as well as Earth and other biosphere-like worlds ... to "purify" and end all suffering? An ultimate triumph of our universe—death!

Audiences never really witness Ghidorah's golden wings in this film. This is because Ghidorah emerges from another universe into ours through black hole singularities, destroying the *Aratrum*. Its three elongated necks descend from the skies above via the singularities—however, only these seemingly disconnected necks are shown ... no body otherwise. Because physical laws in that alternate universe are different from our own, unseen Ghidorah can only manifest in our universe as an expanding gravitational field detected on human instruments. And yet the attacking illusory monster bites savagely into Godzilla. (Viewing audiences can of course see it attacking, but characters in the film do not.) But Metphies is confronted by Haruo (and the spirit of Mothra), and Ghidorah dissipates, sparing Godzilla. Eventually, the remaining space-faring humans' survival is ensured as they band together with the Houtua, in harmony and balance with Nature.

Watching these Toho films, I kept thinking of a 2010s televised post-apocalyptical science fiction series broadcast on the CW Network, *The 100*. This series concerns an outcast band of humans who descend to Earth's surface from an orbiting spaceship sanctuary ring a century after the nuclear war that destroyed civilization to discover whether our planet remains uninhabitable. But they do find sparsely populated human centers, several of which they eventually integrate with before re-launching into space after repeated nuclear attacks devastate the planet once more. Except, in *The 100*, there are no giant monsters, no Godzilla. That is, unless you regard the nuclear detonations throughout as giant monsters, much as Godzilla is reciprocally metaphorical for mankind's folly with atomic and hydrogen bombs.

The pair of Legendary graphic novels sets readers up for impending giant monster apocalypses that came (in 2019 and 2021), while Toho's trilogy shows us a remote aftermath after man has succumbed to the giant monster apocalypse—metaphorical for mankind's unmitigated environmental devastation.[6] There is no battle between a giant ape and Godzilla presented in either set of offerings. The only commonality between these disparate entries is the monster Godzilla (and Ghidorah), while other monster foes are either new to the mythos or, as in Toho's case, old adversaries represented in remarkably different, non-classical form. And yet the theme of utter global degradation caused by mankind, generating giant destructive daikaiju in Medean fashion, pervades through each as it does within the 2010s Legendary filmic series.

CHAPTER ELEVEN

Paleo-Apocalyptical Dino-Monsters

Reflections of Doomsday?

"We cannot command nature except by obeying her."
—Francis Bacon, from his *Novum Organum*
(1620), cited in Greg Keyes's *Godzilla:
King of the Monsters—The Official
Movie Novelization* (2019, p. 123)

How curious that man's preoccupation with dinosaurs correlates with advent of our Anthropocene Epoch.

In its many guises the Prehistoric and Dinosaur Mystique refuses to die and will not as long as our species reigns. Is Man's persistent doppelganger forecasting our future, what is to come—the rites of our inevitable extinction? Man disinterred fossils of prehistoric creatures from within Earth's strata increasingly during the 19th century, and the practice continues with new dinosaurs being described and named scientifically today at an unprecedented, accelerated pace. We may even know more about certain extinct, long-entombed dinosaurian species than we do of far more recent yet long lost ancient archaeological human civilizations buried under shifting sands or enveloped within tendrils of soft jungle moss. And still it may seem that we are hurtling uncontrollably toward some kind of juncture from which the human race may never be permitted to return.

So how, when and why did certain dinosaurs become seemingly apocalyptical? At some point in their anthropomorphic, metaphorical transition and restoration from the rocks, dinosaurs ceased to simply be museum attractions of popularity or scientifically/socioculturally politicized objects, instead increasingly foretelling of festering human plight. In meaning, they became dangerous, foreboding. Because they belonged to the prehistoric past, their anachronistic juxtaposition with man in our present, as experienced in contemporary film and literature, suggested Nature had shifted out of balance. Is a widespread ecological reset needed to restore balance? Driven by our excessive pride and self-confidence (hubris), recesses of our planet intruding into the past, primitive oases within lost places of the globe, let alone our tampering with nature, were all forbidden ... and yet here man—intrepid, foolishly—trespassed anyway.

Our key decade here shall be the 1950s—the period of origin for apocalyptical dino-daikaiju. But let's explore in detail the *before, during,* and *after* the 1950s.

Before the 1950s Assault

For over half a century following their discovery, primeval creatures made fantasy or science fictional excursions ... intrusions and sometimes invasions into modernity. However, it was the dinosaurian interlopers transitioning out of the past that were generally the most popularly regarded of their prehistoric ilk. Initially in the first wave, such genera included the mighty *Iguanodon* and years later, large sauropods such as *Diplodocus, Brachiosaurus* and of course the most famous of the breed then—*Brontosaurus.* Perhaps as expected, yet as properly documented in 2019 by historian of science Ilja Nieuwland, throughout the mid–1800s *Iguanodon* was the most popular of all dinosaurs.[1] Not only was this genus for several decades the most complete, scientifically known dinosaur until then, it also represented a consummation of jolly old England in its golden "heroic age" of paleontology—only to be superseded by discovery of numerous skeletons of the same genus found in a Belgian coal mine during the 1870s (resulting in drastic anatomical alterations in its presumed restoration[2]). Celebrating its stature in the belly of the beast, during its heyday, prominent paleontologists even had a festive publicized dinner seated inside the mold of Benjamin Waterhouse Hawkins's *Iguanodon* model on New Year's Eve 1853 on the Crystal Palace grounds at Sydenham, England.[3] (*Iguanodon's* oft-presumed nemesis, *Megalosaurus*, made a smog-ridden, phantasmagoric appearance at the opening of Charles Dickens's 1851/1852 novel *Bleak House*.)

Relying on the Google Books Ngram Viewer database, Nieuwland who has researched the origin of "dinomania" in Europe determined:

> What immediately becomes obvious when using this tool is that the present dinosaur craze is dwarfed by the mania that took the British Isles during the 1840s and 1850s, after the first dinosaurs (*Megalosaurus* and

Figure 22: Brontosaurus on rampage in London in 1925's *The Lost World*. Willis O'Brien's and Marcel Delgado's genius for stop-motion animation captivated audiences with many such fascinating scenes in this silent movie classic. Today, this city-stomping segment illustrates the hold and psychological appeal of the huge Sauropods during that earlier period in paleontological pop-culture.

Iguanodon) had been uncovered: in the case of *Iguanodon*, its fame was decidedly greater in the United Kingdom than in the United States. ... the construction of ... (Hawkins's) *Iguanodon* model was the *result* of great public interest in prehistoric animals in general, and this one in particular. It did probably help to sustain that interest.[4]

Nieuwland also noted that during England's first dinosaur craze, prehistoric non-dinosaurs were generally neglected relative to the dinosaurs, a circumstance persisting today.[5]

While Jules Verne's 1864/1867 novel *Journey to the Center of the Earth*, introducing the art of "cryptofiction" as some proclaim,[6] was brimming with numerous kinds of prehistoria representing life through geological time—the first full length tale to feature them—dinosaurs, per se, received short shrift. And none of the creatures encountered by intrepid explorers within Earth's deep caverns in Verne's famous novel escaped to the surface where they might plague mankind in one fashion or another. But by the mid–1880s, a curious pairing of prehistoric denizens intruded into civilization for the first time, that is, in an extraordinary fantasy, *visual* sense—illuminating the literary (and later filmic) possibilities! These occurrences first appeared in editions of Flammarion's volume *Le monde avant la creation de l'homme* (1886).

One such peeping Tom (or, rather, "peeping dino") was a rendition of Hawkins's antediluvian *Iguanodon*. Also seen in Flammarion was a curious chimera, an anatomical blend of *Brontosaurus* with *Stegosaurus*, staring intently through an 6th-floor upper skyscraper of the time window.[7] (See Figure 6.) Of the peeping *Iguanodon* image (at one time imagined by Gideon Mantell to have grown anywhere from 70 to 200 feet in length), Nieuwland astutely had this to say:

> The contrast between such huge, unwieldy, and chaotic animals and our own comfortable and controlled surroundings would increase our awe of them (and, of course, our fear). An entire subgenre of "paleo-art" was created to cash in on this confrontation of the ancient with modern life. Predictably, civilization usually suffered the consequences in these depictions. This was, in essence, the artistic version of the "lost world" novel and the predecessor to a long-persisting meme in popular culture.[8]

Certainly during 1900 to the mid–1910s, a crowning of western dinomania fueled by numerous and bounteous fossil bone discoveries made in America, the intriguing nature of the largest of the dinosaurs—the impressive and gargantuan sauropods—usually in the form of *Brontosaurus*, had taken hold on human imagination![9] The creepy and violating, yet still rather tame, ambiance of visuals appearing in Flammarion transcended further into an 1898 portrayal. Here, *Brontosaurus* led the charge ascending to the forefront, peeping through an even higher, 11th-story building window, thus supplanting *Iguanodon's* preeminence.[10]

A few years later, Vincent Lynch's *Scientific American* image of the Gigantosaurus illustrated American Museum of Natural History paleontologist W.D. Matthew's 1914 article.[11] In this striking composition, the mysterious 130-foot-long sauropod dino-monster was no longer simply peeping, but instead nonchalantly striding down a New

York City trolley street flanked by skyscrapers. Matthew supplied a caption for the visual: "If Gigantosaurus *had* lived in the twentieth century instead of many million years ago, if he *could* be made to walk up *Broadway*, New York, the effect *would* be that which we have shown. ... The picture is intended to show how enormous Gigantosaurus was, and is not to be regarded as a representation of his habits." Gigantosaurus was later christened *Brachiosaurus*, a more massive creature (although is now known *not* to have grown 130 feet long), of which a little more shall be said shortly. Such scenes foreshadowed what would soon manifest in a number of dino-monster laden science fiction films, particularly First National's silent 1925 movie *The Lost World* (directed by Harry O. Hoyt).

Although fantasy dino-monsters made their first exploratory incursions into modernity, there has always been a loving, fantastic sci-fi tradition—inspired by Verne—involving intrepid adventurers discovering dinosaurs in isolated, or more generally lost places on, or spectacularly within, the globe, places literally teeming with prehistoric-aspect life, often including huge sauropods.[12] In a few instances, men unwisely made decisions to transport one or more of these prehistoria from their primeval settings into modern cities for display and commercial gain—always with tragic results. Or sometimes, the dino-monsters, otherwise unaided, hauntingly found where we lived ... on their own.

Soon man would wage war with these prehistoric invaders storming into our populated areas via a number of turn-of-the-century works. While the menacing genera were sometimes varied in such tales, a century ago the huge sauropods were often on the vanguard, featured as monstrous therein, usually receiving pride of place over Frankensteinic *Iguanodon*. (And yet still these early incursions should not be misconstrued as apocalyptical in nature.)

Most savvy readers would be familiar with Arthur Conan Doyle's 1912 novel *The Lost World*, in which Professor Challenger's stalwart yet ecologically and anthropologically insensitive expedition conveys a captured pterodactyl to London that escapes through an open window into the night sky. In the 1925 film adaptation, however, a climactic sauropod is substituted for the winged reptile, which smashes a building or two and is fired upon by police before swimming away along the Thames. Yet prior and concomitant to 1912, also the year of the very first dinosaur film—Winsor McCay's animated *Gertie the Dinosaur*[13]—appeared, several early tales involving monsters from the past attacking civilized areas and metropolitan centers emerged, three of which are relatively unknown—two of which involved enraged sauropods. These are Samuel Hopkins Adams's 1903 novel *The Flying Death*, Jules Lermina's novel

Panic in Paris (1910), and an engaging April Fools 1906 *Chicago Sunday Tribune* yarn.[14]

During the early 1990s, following the release of the blockbuster movie *Jurassic Park*, psychologists pondered the question, "Why are children so fascinated with dinosaurs?" leading to a famous essay penned by Stephen J. Gould titled "Dinomania." One answer that materialized addressing so many kinds "from gentle Barney who teaches proper values to young children ... to ferocious monsters who can promote films from 'G' to 'R' ratings" is that dinosaurs are "big, fierce, and extinct ... alluringly scary, but sufficiently safe."[15] However, cultural historian W.J.T. Mitchell cautioned that "at least one thing (is) clear about the dinosaur as metaphor and cultural icon: it cannot be explained by a one-sided characterization as 'big, fierce, and extinct.'" He went on to amend the remark, stating, "The only term we have to summarize these complex and contradictory associations is 'ambivalence,' the tendency to hold opposite feelings, to be 'of two minds' about a single thing.' ... and to pause reflection over a cultural icon whose meaning at first seems completely transparent."[16]

To me, though, the "big, fierce and extinct" formula seems informative or at least sufficiently instructive, that is in relation to understanding the cultural evolution of filmic and science fictional dino-monster derivatives stemming from real dinosaurs known to science throughout their historical reappearance and symbolic revivification on Earth. In Godzilla's specific case, or others sometimes regarded as apocalyptical creatures, another key feature or characteristic adding to their popularity and consumer fascination—*radioactive*—complements the formula. Thus, parameters of big, fierce, extinct (i.e., or supposedly so) and radioactive may well define the foundational, classic dino-monsters of science fiction and horror as known to and preferred by modern audiences particularly of the 1950s and consumers living in our eco-cidal, apocalyptical Anthropocene Era.[17]

Let's further explore the "big" aspect of this formula. As Nieuwland noted, the late 19th and early 20th centuries were marked by a philosophy that "bigger was always better ... a fascination for hugeness, be it skyscrapers, bridges, zeppelins, or ocean liners. Large size was equaled with modernity, with progress, and with importance ... the biggest guns ... the fastest and largest ships. ... And (Andrew) Carnegie was soon to get his 'most colossal' animal." Further, it was an age in which "size truly mattered, being the biggest remained important."[18] An apotheosis of the age, New York's Empire State Building was constructed in 1931, and was mounted—*conquered*—by an even more spectacular giant ape—Kong—who had already scored victory over much larger-than-life prehistoric monsters on his own Skull Island. Of course, another factor favoring the sauropods (i.e., *Brontosaurus* and *Diplodocus*) a century ago was the fact that their bones (or plaster replicas thereof) were among the first to be assembled in museum mounts for public edification.

As I stated in a 2017 *G-Fan* article: "Universal appeal of potentially menacing dinosaurian BIG-ness truly expanded with discovery and announcement of the largest of the clan—the great long-necked sauropods, of which spectacular specimens were disinterred and 'resurrected' with great hoopla in North America and Africa between 1877 and 1915! Following their scientific description, as disseminated into the mainstream, the pop-cultural era of the HUGE dino-monster had begun."[19] "Big" was clearly in then, and arguably the acme of pop-cultural sauropod predominance then was attained by the early 1930s through the Sinclair Oil Refining Company's gasoline brand logo, wherein a brontosaur seen in lateral view was adopted.

In 2018 I directed a savvy-minded panel of giant monster movie lovers to present our investigation of the "top ten sci-fi prehistoric monsters" at the annual Godzilla convention held in Chicago known as G-Fest. Particulars of how we defined "prehistoric" and other important analysis particulars are to be found in a *G-Fan* article I subsequently wrote.[20] Godzilla and Kong tied for first place in our categorically organized rubric (*not* determined upon a simple survey or panel/audience vote). Yet, out of a field of 47 monster nominees (archetypal *Iguanodon* did not make this initial cut), venerable *Brontosaurus* made our top ten list—finishing in a tie for 8th place! Our panel of experts absolutely recognized this dino-monster's appearances in a number of films (especially perhaps 1925's *The Lost World* and 1933's *King Kong*), and published stories, therefore registering its overall importance to the genre at large. But *Brontosaurus* wasn't the only major sauropod known to the public then, as this was the period when newspapers often reported intriguing stories concerning the next "most gigantic fossil bones," or even allegedly living prehistoria witnessed by man ... and so on.

Three kinds (genera) of dinosaurs principally became associated with the public's perception of sheer, utter reptilian massiveness: *Brontosaurus*, *Diplodocus*, and *Gigantosaurus* (later named *Brachiosaurus*). Nieuwland outlined the 1904/1905 pop-cultural battle waged between *Diplodocus* (of Pittsburgh's Carnegie Museum of Natural History), and New York's American Museum's *Brontosaurus*. Both institutions were in the throes of erecting the world's first skeletal mount of a sauropod; the rewards were public (as well as personal) adoration as well as permanency upon the human psyche. Scoring early points was the Carnegie, which built a temporary trial mount of a cast of their *Diplodocus* in the exhibition hall of the Western Pennsylvania Exposition Society on June 29, 1904. The skeleton was removed by July 2, however, after which the plaster replica was shipped to London where it would be placed on display by May 12, 1905, in the British Museum (Natural History). Carnegie's original skeleton—from which several plaster casts were made for distribution—was eventually displayed in Pittsburgh on April 11, 1907.

Of this competition, it is said, the Carnegie's paleontologist "William Holland was particularly eager to contrast *Diplodocus* with the New York *Brontosaurus* that was supposed to be unveiled soon. In a letter to (Andrew) Carnegie, he took pleasure in denigrating the American Museum of Natural History's effort, calling it 'an object in plaster of Paris,' and remarking casually (and quite untruthfully) that 'we have enough material to set up a skeleton of *Brontosaurus* ... a clumsier beast than *Diplodocus* and not nearly as long."[21] So, which institution "won" this early competition of sauropoda? Certainly, a century ago, *Diplodocus* held its lofty household rank and polysemous, metaphoric role in popular culture, yet, by the 1920s, was clearly eclipsed by *Brontosaurus*. While the American Museum's skeleton, referred to by Holland (1848–1932) as a "caricature,"[22] did contain many plaster bones, most of it was genuine fossil. Meanwhile, detractors of the first Carnegie replica took note of the fact that because their *Diplodocus* housed in London was entirely plaster, it was merely an "imitation dinosaur."[23] Nonetheless, the American Museum's *Brontosaurus*—the world's first permanent sauropod display (great grandfather of later filmic dino-daikaiju), which turned out to be a most influential exhibit for the future of dino-monster-dom—was dutifully on display by February 1905.[24]

It remains possible that general public fascination with sauropods—*Brontosaurus* in particular—stemmed from additional visual influences apart from institutional displays of these gigantic brutes that were going up in museums around the world, published newspaper and magazine accounts and reproductions of painted restorations, for example. But it seems rather clear that the American Museum's brontosaur display, with other exhibit enhancements of 1905, set the stage for three imaginative pioneers of movie-land's special effects industry who created groundbreaking realistic scenes of dinosaurs interacting with man between 1912 and 1933. The foundational filmic artistry of Winsor McCay, Herbert M. Dawley and Willis O'Brien, who were mentioned in Chapter Two as filmic paleo-pioneers, was stimulated by the American Museum brontosaur (i.e., not a *Diplodocus*) display, leading to widespread seminal comprehension of how dinosaurs were supposed to move and live in their prehistoric settings, at least in popular view.[25] Merkl suggests that McCay's "Gertie" may also have had a sort of subliminal influence upon Merian C. Cooper, who went on to *King Kong* fame with RKO.[26] However, such evidence would seem generally lacking, not only in regard to Kong but also with respect to Godzilla.[27]

For decades *Brachiosaurus* seemed to be the dinosaurian record holder with respect to mass and size, but records are always set to be broken and so into the new millennium, the current record-smasher is a South American sauropod known as *Argentinosaurus*. While *Brachiosaurus* had been described in 1903 from fossils excavated in Colorado in 1900, the most famous skeletal mount was constructed from fossils found during 1908 to 1914, at Tendaguru, Tanzania. A stop-motion animated *Brachiosaurus* even appeared shortly in *The Lost World's* dinosaur stampede, a marvelously orchestrated scene in which a multitude of dinosaurs flee an volcano erupting on Professor Challenger's lost plateau. Amidst the throes of two world wars, that specimen was finally built and displayed in Berlin in 1937. The Berlin (African) specimen, originally referred to as *Gigantosaurus*, then later as *Brachiosaurus*, is now known as *Giraffititan*. Still, even the largest of real sauropods known to science would appear puny in size standing next to 1954's 150-foot-tall Godzilla.

While in Verne's 1864 inaugural effort the explorers never went on the warpath hunting or attacking primeval creatures they encountered, Henry Francis's 1908 short story represents one of the earliest in which modern humans fired upon and killed a prehistoric endangered crypto-dinosaur—in this case belonging to the sauropod clan.[28] Such a bloody, hell-fire trend would only escalate in coming decades, as living dinosaurs and other (endangered-last-of-their-kind) prehistoric denizens would be increasingly regarded as menaces, science fictional props worthy of fighting and destroying using guns and ammo.[29]

It is not a stretch to claim that by the early 1930s, sauropods and particularly emblematic *Brontosaurus* had supplanted *Iguanodon* as the most symbolic if not popular of the dinosaurian clan, most assuredly due to their affirmed great sizes attained in life. But this infatuation with the hugeness of prehistoric animals would soon metamorphose into a fascination with dinosaurian *ferocity*, a trait that could not be sustained by the gentler sauropods. By 1934 the Sinclair Refining Company's famous brontosaur logo symbolized "the vast age of the crude oils which are refined into Sinclair lubricants."[30] Essentially through pervasive advertising, venerable *Brontosaurus* had instead become relegated to a commercial marketing icon forsaking its lead position in symbolizing paleontological might and dread. Another dino-monster more imposing in its distinctively fierce persona would soon fill the vacated pop-cultural niche! True, some science fictional brontosaurs of the time were cast as vicious or even as monstrous flesh-eaters—however, by the mid–1930s those enormous creatures had given way to another iconic pop-cultural dinosaurian form, the theropod breed. From then on, prehistoric world fierceness would most legitimately and convincingly projected through *Tyrannosaurus rex's* terrifying visage.

Arguably, North America's *Tyrannosaurus* became predominant during the early/mid–1930s especially due to three influencing pop-cultural factors. First was Knight's Chicago Natural History Museum Late Cretaceous mural, showing this genus facing off against an advancing three-horn *Triceratops*—perhaps second only to Zallinger's magnificent *Age of Reptiles* painting as the foremost portrayal on canvas of any prehistoric scene or distinctive setting. Secondly, two animatronic/mechanical life-sized and lifelike *T. rexes* (e.g., along with *Triceratops*) made appearances at the Chicago 1933/34 World's Fair, both at Sinclair's touted exhibit and also within the Messmore and Damon "World a Million Years Ago" display. And thirdly, *Tyrannosaurus* lost a major title bout for the ages witnessed by millions all over the globe against none other than Kong himself on Skull Island in RKO's 1933 blockbuster film *King Kong*. Nearly nine decades afterward, *Tyrannosaurus* finished a respectable 3rd in my 2018 "sci-fi prehistoric monsters" G-Fest panel. So now let's consider the "fierce" aspect of the psychologist's equation, as we approach the apocalyptical stage of dino-monster and daikaiju history.

Like all real dinosaurs and famed paleontological monsters of film, *Tyrannosaurus rex* has evolved both in scientific restoration as well as metaphorically in popular culture. As mentioned in Chapter Five, Rex's history can be comprehended in three stages—savage, lordly, and renaissance. Prior to 1950, Rex was generally noted as the ultimate savage king, reigning when reptiles were lords of creation, perhaps most representative of a Darwinian nature-red-in-tooth-and-claw existence that so many assigned to our primeval world. Besides a trio of Knight's magnificent paintings,[31] we see this image projected through filmic scenes such as Kong's 1933 battle with the tyrannosaur on Skull Island, Disney's 1940 animated stormy combat in *Fantasia* ("Rites of Spring" segment directed by Paul Satterfield) between a doomed stegosaur versus a vicious tyrannosaur, and even in Max Fleischer's directed 1942 cartoon Superman episode—"The Arctic Giant," the latter case involving a pseudo-tyrannosaur slugging it out in a city center, Metropolis.[32] But it is the later lordly kind (i.e., Zallinger's) that lent its bipedal, carnivorous form to the original Godzilla costume. Today we often see artistic renditions in books and magazines of the 1970s and beyond, Renaissance Rex adorned with feathers.

Nevertheless, all three versions of Rex were each in their heyday generally considered as imbued with the Mesozoic's most fierce, blood-thirsty persona and disposition. Fueled by horrors of two world wars, the appeal of the wayward, bewildered sauropod lost or misplaced in modernity eventually became preempted by the ferocious bestial pseudo-theropod; rapaciously ravaging formidable jungle prey, destroyer of cities, emblematic of pending global warfare. Arguably, however, the heyday of the utterly soulless fierce dino-monster didn't materialize until the Cold War's onset.[33]

During the 1950s Paleo-pocalyptical Assault— Specter of Extinction

Prior to 1940, mankind was afflicted with an overweening confidence in his capability to dominate, exploit and control Nature. Through the late 1940s, reeling in the horror and wake of two world wars—ending in a pairing of (fission-fueled) atomic bombings—a global pandemic (i.e., of 1918), and while in throes of America's Great Depression, a decades-long period of woe that certainly must have seemed socio- and geo-politically apocalyptical to many, prehistoric life (which generally fell into neglect during this time) and associated legions of science fictional dinosaurian invaders hadn't been regarded quite as such.

So, what happened during the early 1950s that led to dinosaurs becoming metaphorical monsters of Armageddon in popular culture? The short answer is the Soviet Union's development of their fusion hydrogen bomb (tested in 1953), vastly more powerful than the atomic bomb, to counter our own thermonuclear invention (first tested in 1952), thus heightening nuclear fear and increasing anxieties over the human race's and civilization's capability of surviving a next or possibly final war.[34] Over-confidence in man's technological ability to conquer and impose our sense of order on natural systems, and in so doing direct his destiny, waned.

Would mankind go the route of the dinosaur ... not after flourishing for their stout 140-million-year reign, but regrettably all too soon, in a momentary flash? Mankind's extinction increasingly seemed not only inevitable, but palpable. Despite the perceived might of any clade, according to Darwinian law no species was anointed or indestructible.

Invoking the psychologists' threefold criterion over popular dinosaurian intrigue, the fact that such fascinating big or fierce creatures as dinosaurs *were* extinct added a certain urgency to the fact that someday (all too soon, perhaps) it may be our turn as well to be administered rites of extinction. Big, fierce, extinct—through his own machinations, puny man was on the verge of becoming a dinosaur. Dinosaurs had become paleo-poster children of the extinct, confined to their Mesozoic segment of Zallinger's famous mural. It was difficult to think of them without pausing to reflect upon when in the context of Time they once dwelled ... and died. To a large extent, their signifying purpose (as *the* most popular set of extinct organisms felled by Nature) was to reflect on geological time and extinction—the sobering fact that it does indeed happen, and that it is pending for all life.

The science fictional seeds of doomsday had been gathering for a while, especially (for our purposes here) since the publication of H.G. Wells's 1914 novel *The World Set Free*, in which the illustrious, visionary author contemplated warfare use of an atomic bomb (that would "continue to explode indefinitely" in a chain reaction) delivered from attacking airplanes.[35] Curiously, dino-monsters and other prehistoria, swarming from caverns deep inside the Earth, had then made an early literary association with radiation, per Lermina's *Panic in Paris* novel, or later in film via 1951's *The Lost Continent* (Lippert Pictures, directed by Samuel Newfield), in which for the first time menacing dinosaurs were also associated with atomic missiles and radiation.[36] But now paleontological creatures that should rightfully be extinct—creatures we exhumed from rocky strata—underscoring Nature's imbalance, seemed poised to strike an early Gaian (or Medean) death blow out of the deep past, from geographically lost regions of, or from within the bowels of, our planet onto our own turf, to avenge our misguided ways.

The public increasingly feared the prospect of a world war from the late 19th century onward, in part because of the newly refined warfare inventions—weapons of mass destruction—devised by military engineers and scientists. Historian of science and biologist Peter J. Bowler noted that during the 1920s (in the wake of a world war and a pandemic), public fears escalated over the possibility of planes carrying and delivering "death germs" spewed from above to populated areas. Wells had written of futuristic, science fictional forms of aerial warfare, and Winston Churchill echoed in 1932 that such attacks spewing deadly biological agents were possible. According to Bowler, "the terrors (then) imagined ... were fully as comprehensive as those depicted in fictional accounts of a nuclear holocaust ... during the (later) Cold War."[37] Under contemporary terms, the 1930s public reaction to the prospect of

Figure 23: Lisa Debus's 2021 recreation of a scene showing an imposing, yet rather nondescript, prehistoric monster emerging from a subterranean environ beneath Paris, as conveyed on the cover of Jules Lermina's 1910 novel *L'effrayante aventure* (translated as an English edition in 2009— *Panic in Paris*). Lermina's is one of the first stories to connect a dino-monster invasion with radioactivity (used with permission).

disease germs raining from the sky was to citizens of the late 1950s the equivalent to a radioactive ash cloud.

Most would associate plot-wise freeing from frozen suspended animation of the stop-motion animated dino-monster named "Rhedosaurus" in Warner Brothers' *The Beast from 20,000 Fathoms* (directed by Eugene Lourie, and released on June 13, 1953—just two months prior to the test-detonation of the Soviet Union's first hydrogen bomb) with then contemporary nuclear angst. The movie opens with scientists conducting "Operation Experiment," although they're arguably triggering a test thermonuclear device in the frigid Arctic. However, the revivified movie monster's chief menace stems not so much from radiation exposures harmful to humans or its fierce, carnivorous, building-smashing tendencies, but instead from the long dormant paleo-lithic germs—to which modern man has no immunity—contained in its blood,

Figure 24: Decades before their co-production of *Godzilla vs. Kong* with Legendary, Warner Brothers released 1953's *The Beast from 20,000 Fathoms*, with its stop-motion animated Rhedosaurus's viral super-spreader event in New York, causing people to sicken and die from the ravaging dino-monster's tainted blood. Many film historians believe *Beast* was a direct precursor to Toho's *Gojira*, released the following year (author's collection).

threatening a biological plague![38] In time, sickness caused by exposure to radiation would become more understood by physicians and medical scientists (especially those practicing in Japan[39]), but messaging in 1953's *Beast* straddles the historical period between when death germs were feared as much or more greatly so than risks of exposure to radiation fallout.[40] *Beast's* bio-terror theme was instead probably conceived as just a cool sci-fi, natural consequence of digging into a forbidden past, ironically facilitated with (then) futuristic nuclear weaponry via "Operation Experiment."

After 1953, attitudes quickly shifted. As Spencer R. Weart stated, "It all began with dust. Already in planning the Trinity bomb test the Los Alamos scientists had worried about the dust that the explosion would hurl into the air. Passing through the swarm of neutrons and fission fragments in the fireball, the dust would become dangerously radioactive, then drift downward." And what was perceived as a more localized yet still dangerously radioactive dust fallout in the case of an atomic bomb would be of an exponentially worse magnitude in the matter of a thermonuclear hydrogen bomb detonation! Weart further noted, "It turned out that a hydrogen bomb could kill people not only nearby, but *hundreds of miles away* ... releasing nuclear energy was a blasphemous violation of the entire planet"[41] (my italics).

Sci-fi film maven Bill Warren astutely stated that in 1950s sci-fi flicks, the radioactive gimmick wasn't significant. Warren claims, "although the 1950s did tend to be worried about atomic warfare, radiation in SF films wasn't a means of expressing this fear, probably not even unconsciously. It was just a way of originating an unusual or interesting menace, or explaining one already conceived. Radiation was used to explain many wonderful things, from giant insects to walking trees to resurrecting the dead. This was not a form of nuclear paranoia, merely cheap and simple plotting."[42]

A problem here is that while Warren is generally correct for most B-movie cheapies, downplaying radiation's significance on public paranoia in a number of other important films, but is clearly wrong in the grim case of 1954's *Gojira*, where the hydrogen bomb threat theme is pivotal, nearly palpable. Consider that memorable scene, for example, on Odo Island, when Dr. Yamane observes a crackling Geiger counter-analysis within Godzilla's footprint; a radioactive trilobite, until then known only in the fossil record, is found embedded in one of these prints thus linking radiation with the inevitable. In Godzilla's visage, radiation is an unnatural manifestation of man's inhumanity.

Spencer Weart noted how a dawning *environmental movement* (in America) stemmed from growing contemporary concerns over radiation poisoning via nuclear bomb ash fallout. For he stated in 1988, "There were many reasons for the rise of the environmental movement, but the basic themes—dismay with technological authorities and systems that seemed about to doom the entire world—had first been thrust upon the public by hydrogen bombs."[43] Weart further credits language used by Rachel Carson in her influential 1962 book *Silent Spring*, bridging nuclear and chemical pollution/environmental themes and imagery. For Carson wrote of an imaginary USA town, presumably resultant of pesticide crop dusting.[44] Clearly, nuclear fallout imagery—steeped in looming fear of the hydrogen bomb, our Sword of Damocles—was used to emphasize a more chemically-driven phase of the environmental movement.

So perhaps, by mid–20th century, it became fitting and conventional for fossil creatures from our geological past—dinosaurs and pseudo-dinosaurians alike—to symbolically warn mankind of our polluting "human volcano,"[45] our current and persistent folly in recklessly wielding industry and technology. During that same year when Isaac Asimov's precociously-minded paper was published (see note 40), Universal Pictures released *Creature from the Black Lagoon* (directed by Jack Arnold, 1954), incorporating an inaugural scene symbolizing Man's unyielding pollution of the environment. For not only does "Kay," played by beautiful Julia Adams (1926–2019), flick her cigarette butt into the Devonian Gill Man's natural habitat (i.e., *his* Black Lagoon), but also male scientists aboard the *Rita* trawler haphazardly poison the waters with a toxic chemical—rotenone—in order to capture the prehistoric living yet (like the doomed sauropod in Henry Francis's 1908 story) endangered fossil crypto-specimen. By decade's end, the Soviet's successful launching of rockets and spacecraft into orbit further forecasted doom: America was losing the space race. Instead of "death germs" raining from the sky, would there instead be hydrogen bombs descending upon us from outer space? Man's environment and that of the natural planet we presided over became increasingly under fire in real world America as well in reel-world monster-dom throughout the 1960s and beyond.

After the 1950s—Paleo-pocalyptical Repercussions

The cultural age of dinosaurs—a term used loosely here—throughout their history as discovered and comprehended by man may be divided into three interwoven phases.[46] In the earliest, dinosaurs and collectively other prehistoric forms of life known to the masses traditionally told of (or symbolized) Earth's deep past and those periods of the geological time table in which they lived. By the 1950s, as we've just outlined, dino-monsters from deep time (dino-daikaiju such as Godzilla, Gigantis, Anguirus, Rhedosaurus, Paleosaurus, Rodan, Reptilicus, Gorgo and Korea's Yongary) increasingly warned mankind of the possibility, if not inevitability, of extinction—especially given the direction humans were heading—that would result from either natural causes, or through our own insidious devices. It's as if they grinned through toothy skeletal smiles, "if this goes on then welcome to our Apocalypse."

During that second phase, most distinctively during the 1950s through the very early 1960s, such warnings from dino-monsters and invading prehistoria were usually of a suggested metaphoric, read-between-the-lines-for-full-implications variety. Beginning during the 1950s yet increasingly so by the 1980s, however, messaging forecasting potential doom became more direct, even verbal, in-our-faces.[47] "Get a grip, Man—think and look at what you're doing!" their all-too-familiar visages cried. Now intelligent dinosaurs while increasingly assuming *anthropomorphic* guises in the form of "dinosauroids," a science fictional variety of dinosaur that hardly existed in our annals prior to a time when scientific intrigue intensified over how quickly and why dinosaurs were extinguished, directly warned us.[48] Through their (usually novelized) actions, mimicking our own, we should have been able to learn how to avert doomsday. Would we?

Beyond the 1980s, dino-monsters symbolized other potential forms of chaos and annihilation (besides the ever-present radiation threat[49]) that mankind would suffer upon himself and Earth through his hubris, wielding technology in an ill-advised manner. Public fear cascaded into other realms and possibilities: tampering with biotech, genes and microbes especially, loomed at the heart of the matter. The 1953 Rhedosaurus's deadly strain of pathogen contained in its blood droplets was not weaponized, and yet was an inadvertent outcome of unwisely experimenting with the atom. By the 21st century, science fiction writers were already recruiting prehistoric genera in gruesome mass mortality biological warfare, in which paleo-prions and germs preserved in an ancient, once isolated Antarctic lake (Lake Vostok) infect and produce zombified dino-monsters in a novel trilogy launched in 2014's *Jurassic Dead*.[50] The chief bio-terrorist under this "new world order" orchestrating the horrific global pandemic onslaught intends to become a "god."[51]

As we learned in Crichton's *Jurassic Park* (1990), genetic power trumps atomic power, and neither should be wielded without restraint (lest, akin to that Victorian age dinner inside the belly of Hawkins's *Iguanodon*, we get consumed metaphorically by creatures that expired millions of years ago).[52] Through his vicious tyrannosaurs and velociraptors, Crichton showed us indelibly how and why we should fear lurking dangers of biological engineering. But Crichton's is not the first modern tale illustrating consequences of such rash conduct. The origins of the mad scientist's unscrupulous use of biotech reach back to Mary Shelley's *Frankenstein*, and are later echoed in Wells's novel *The Island of Dr. Moreau* and Aldous Huxley's *Brave New World*.[53] A 1928 story published a few years before release of Universal's classic film *Frankenstein*, told of the "resurrection" of a Siberian Mammoth using galvanic currents and biogenic rays, causing the specimen to revive.[54] Sci-fi revivification of dead or long-dormant prehistoric life through one technological means or another[55] has become a standard means of accomplishing the impossible—confronting man with extinguished, yet living monsters out of time's abyss, a past that remains as forbidden as his future portends ... if this goes on.

During the new millennium, kaiju-writing became a passion for many writers, now aided by an array of independent publishing platforms like Createspace and Kindle. Perhaps the foremost author who has scored the most success in pseudo-dino-daikaiju fiction writing, however, is Jeremy Robinson, whose *Project Nemesis* (2012) ranks as the "highest selling original (non-licensed) kaiju novel of all time."[56] The monster in this gripping novel, carried forward through three additional sequels is "Nemesis," engendered without need of irradiative boost, nor fused from dinosaurian genetic stock as in Crichton's novels, but rather from a hybrid merging of alien with human-female DNA. Yes, troubles with misuse of genetics abound in Robinson's series of novels. Nemesis rapidly metamorphoses through four stages from a human-sized creature (rather akin to Toho's *Shin Godzilla)* into a towering city-smashing Godzillean monster replete with vertebral spines and a long thorny slashing tail, and capable of spewing fiery gelatinous material from glowing furnaces on her abdomen, blasting everything for surrounding miles. And to make matters far worse, as in Robinson's novel sequels, the evil genius behind Nemesis's dark genesis also generates a number of other genetic daikaiju monstrosities to further plague mankind.[57]

So, who are the true monsters in such fictionalized yet metaphorical stories of ill-advised biotech, tissues and genetics gone awry? Have we learned nothing from over two centuries following Mary Shelley's groundbreaking novel? Isn't that monster visible in the mirror, as Shelley once illustrated and Robinson suggests? Hubris—man playing

God—does it ever really play out well? Goddess Gaia may seek her revenge.

During Ascendancy of Apocalyptical Dino-Daikaiju Time

In 2018 Bill Gates stated, "Of all the bad things that could happen—a nuclear war, an asteroid, a gigantic earthquake—the one that's most scary is a big epidemic, like a flu epidemic sweeping the world as it did in 1918."[58] Yes, dangerously monstrous mutations may occur from undue radiation exposures, or even perhaps microbially within our own host bodies following super-spreader viral events. How prophetic for what ensued globally in 2020 through the time of this writing. Meanwhile, in 2020, news that China was conducting genetic experiments to develop "super soldiers" hit the airwaves. Later, on January 27, 2021, a reporter on MSNBC Live television network announced breaking news that the *Bulletin of the Atomic Scientists* had moved the dial on the Doomsday Clock up to "100 seconds before midnight."

No wonder, just to chill from it all, we demand entertaining escapism at the movies (or on streaming services) and page-turning novels in order to lift dour spirits in spite of impending apocalypses!

Gaia, Kong and Godzilla—which entity will ultimately prevail?

Epilogue

C'mon. So What?—
You Can Never Go Back.

"The extraordinary, rapid growth of the Homo sapiens population, coupled with its voracious appetite for planetary dominance and resource consumption, had put every measureable biological and chemical system on earth in a state of imbalance."
> —Laurie Garrett, *The Coming Plague*,
> Penguin Books, 1994, p. 550

"Most people think the world is gonna end in a ball of fire, or a blinding flash—a giant mushroom cloud. But ... I think Armageddon is happening all around us, in slow motion."
> —"Aitor Quantic" (Richard Gunn) remarking
> on the Anthropocene Era in *Hemlock Grove*,
> Season 3, Episode 5 (Netflix, 2015)

"Man can hardly recognize the devils of his own creation."
> —Dr. Albert Schweitzer quoted in
> Rachel Carson's *Silent Spring*, 1962, p. 17

"In philosophical terms, the future is, after all, a vulnerable and developing present."
> —Senator Al Gore in *Earth in the Balance:
> Ecology and the Human Spirit*, Houghton
> Mifflin Company, 1992, p. 235

169

*"The end is the beginning. ... The world goes to sh*t;
everybody's a philosopher."*
—AMC's *Fear the Walking Dead*, Season 6,
Episode 3, first aired on 10/26/20

Challenges we face! Man versus Nature, individual rights and freedom versus collective objectivity ... with civilization preciously hanging in the balance. And prehistoric apes versus metaphorical dino-monsters out of geological time. Kong and Godzilla. ... Gaia against Medea—how do such matters interrelate and for our present purposes here, what does it all mean? As movie director Shin'ya Tsukamoto mentioned in a 2009 televised film *Monsterland* (directed by Jorg Butgereit), monsters are symbolic threats to and from Nature. Monsters manifest in order to teach us important lessons. A stark message for the 21st century (and beyond) is pending environmental apocalypse as explored in filmic and literary portrayals, unfolding when or ... *if this goes on.* Fearful and foreboding socio-political extrapolations concomitant with science fictional suspension of disbelief warn ominously of Earth's revenge—our eventual eradication.

Whereas in Japan, audiences embraced prominent environmental/ecological themes through a six-decade history of *kaiju eiga* "message" films, during the 2010s (and into 2021) Legendary employed alternate, less politically charged approaches to portraying such topics, beginning with 2014's *Godzilla*. Rhoads and McCorkle state this "is not to say there are no environmental issues relevant to America, Japan, or the world as a whole in the film (i.e., 2014's *Godzilla*), but it seems that Hollywood eschewed any kind of political undertones that might estrange possible audiences at home or abroad."[1]

So rather than overtly heaping loads of reactive self-blame for our history of nuclear proliferation and politically charged anxiety over (ever-worsening) climate change—i.e., global warming concomitant with diminishing thicknesses of sea ice and glacial ice coverage, increasingly chaotic weather patterns, and extreme or unusual weather events—in the new millennium, in this case Hollywood demurred. Legendary with Warner Brothers suggestively resorted to exploring the idea of Godzilla and Kong as potential (possibly reluctant) saviors of mankind in throes of, or against, a wrathful out-of-balance Mother Earth deified via its Hollow Earth womb.

Throughout the ensuing series, Gaia/Medea—although unnamed as such—deploys vengeful monstrous cohorts (e.g., MUTOs, King Ghidorah, Rodan, Skull Crawlers) resuscitated out of the subterranean bowels of the planet, allied with like-minded individuals intent on making the planet green at any cost and in harmony once again through extermination of excess human population. These largely self-serving eco-terrorists seek recrimination against civilization for our steadfast polluting ways, so toxic to biota and fragile ecosystems. Why go to such an extreme? As some do believe, "Extrapolations of current trends in the reduction of diversity imply a denouement for civilization within the next 100 years comparable to nuclear winter."[2]

Although cellular mitosis has been ongoing for over a billion years, no species lives forever, seemingly with the exception of those nastily persistent cockroaches (thriving since the Coal Age, Carboniferous), and rather than gradually fading away through time, often species are wiped out in a symbolic clock-of-doom geological instant. Is that our (all too sudden) foreboding fate presaged by Legendary's and Warner Brothers' symbolic rebooting of Kong and Godzilla, with enraged Mother Earth poised to give us that ultimate heave-ho, sooner than later?

To be sure, there's an air of deep nostalgia in writing this ode to an old monster movie held so dearly in childhood. Along with dinosaurs, I grew up with those anachronistic prehistoric monsters Godzilla and Kong during the late 1950s and early 1960s. I cherished these monsters. Yet to be exposed to 1956's *Godzilla, King of the Monsters* on Chicago-land television during the midst of the grim Cuban Missile Crisis (while I was blissfully unaware of Rachel Carson's ecological researches) was harrowing indeed, given—that October 1962—what would have happened if Fidel Castro gave the launch order.

Yes, there were a slew of raging filmic radiation monsters plaguing mankind then, but to me unnatural Godzilla reigned chief among them. Kong (never one of those irradiated) was his greatest adversary, and it seemed that—in 1963—natural, evolutionarily ancestral Kong represented our salvation if he won (which appeared to many then to have been the case). But would that become humanity's redemption, or just another USA (symbolic and heroic) military-style victory? (And yet, perhaps like young Sam Brody, wasn't I secretly hoping the dinosaur would win?)

As early as 1959, Dr. David Price of the United States Public Health Service warned, "We all live under the haunting fear that something may corrupt the environment to the point where man joins the dinosaurs as an obsolete form of life."[3] Chemical pollution doesn't always manifest as illness upon initial exposure: as in the case of certain deadly viruses, many years may transpire before development of symptoms. As Rachel Carson stated in 1962, therefore, "It is human nature to shrug off what may seem to us a vague threat of future disaster."[4] That remains a primary reason for our tardy response to planetary climate change.

God's wrath or simply predictable results of our overweening sense of technological superiority? Regardless, we're all trapped in the same fragile and evolving biosphere, with nowhere else to go. Chief Seattle stated in 1855, "the Earth does not belong to man, man belongs to the Earth. ... Man did not weave the web of life, he is merely a strand of it. Whatever he does to the web, he does to himself."[5] Giant monsters may beckon and warn us, but as the battle rages, Earth herself (Gaia? or Medea?) will ultimately decide our fate. Who or what will punish more? If monster movies are about insults to nature—the web of life—then are we ultimately the monster?

Espousing the then developing western view of modern environmentalism, Rachel Carson went on to state in her *Silent Spring* (1962):

> It is fashionable to discuss the balance of nature as a state of affairs that prevailed in an earlier, simpler world—a state that has now been so thoroughly upset that we might as well forget it. Some find this a convenient assumption, but as a chart for a course of action it is highly dangerous. The balance of nature is not the same today as in Pleistocene times, but it is still there: a complex, precise and highly integrated system of relationships between living things which cannot safely be ignored any more than the law of gravity. ... The "control of nature" is a phrase conceived in arrogance, born of the Neanderthal age of biology and philosophy, when it was supposed that nature exists for the convenience of man. The concepts and practices of applied entomology for the most part date from that Stone Age of science.[6]

Japan's 20th-Century Green Giant Monster Films

Let's consider Japan's evolving movie industry take on giant monster (e.g., kaiju) films addressing perspectives on natural, "green," and environmental themes released before Legendary's new-millennial contrasting reexamination of the general theme. Of course, these older films produced by Toho and Daiei abound with daikaiju, although conveying far more ideologically than just a bevy of giant wrestling, battling beasts. Most critical will be discussions of Toho's *Son of Godzilla* (1969), 1971's *Godzilla vs. Hedorah* (U.S. release in 1972 as *Godzilla vs. The Smog Monster*), Daiei's *Gamera vs. Zigra* (1971), *Godzilla vs. Biollante* (1989), and *Godzilla vs. Mothra* (1992). One other, later Toho film, 2016's *Shin Godzilla*, where the titular monster is itself viewed as an evolving Hedorah-like pollution creature, must also be examined.

Most western fans of giant Japanese movie monsters are familiar with the traditional threat-of-radiation theme, but less so with other films conveying dangers posed by pervading chemical pollution—Smog

Monster's acknowledged notoriety notwithstanding. And most monster movie fans might be even far less familiar with Japanese filmic monsters symbolic of global climate alterations resultant of science gone mad, exacerbated by concomitant anthropogenic waste dispersal.

Al Gore reflected how, during the ice ages, "What is now New York was covered by one kilometer of ice, even though world temperatures were only 6 degrees Celsius colder than today. If such a small change on the cold side caused the ice ages, what can we expect from a change of that size on the warm side? Moreover, while those changes took place over thousands of years, the ones now predicted may occur in a single lifetime. ... incredibly disruptive to civilization."[7] In *Son of Godzilla*, directed by Jun Fukuda, originally released in Japan in 1967 and later to American television in 1969 by the Walter Reade Organization, experimentation is underway on a South Pacific isle to transform and arrogantly control Earth's weather system and climate ... addressing the famous Malthusian problem for the good of mankind.[8] But as we know, what's considered good for man may not be optimal for the planet! Still, sometimes without calculated risk, there may be no reward.

When an atmospheric radioactive dosing conducted over Solgell Island goes awry, the local climate suddenly is converted into a hellish, 160-degree neo-primeval "steamy inferno."[9] This environmental perturbation rapidly causes mutations: bugs, praying mantises and spiders hatch from egg casings at gigantic sizes. Minya also hatches during this tumultuous occurrence, only to be attacked by the bugs. But his anthropomorphized parent Godzilla arrives to save the day, destroying the huge bugs more effectively than Terminix ever could.

Son of Godzilla is often derided for its introduction of, well, Godzilla's rather dorky looking and comically odd-behaving son, named Minya or Minilla. Consequently, the precociously relevant messaging about our tampering (scientifically, industrially or otherwise) with the environment is usually ignored. While scientist Dr. Kusumi (Tadao Takashima) claims their actions are benevolent, he also realizes that if their technology got into the wrong hands (or countries), terrorists "could also freeze the entire world. The results would be the same as a nuclear holocaust." When wielding science and technology, it's unwise to calculate poorly.

As Bogue aptly noted, an association between development of a weather-altering radioactively charged system with a potential planetary-freezing event if the technology were to be abused presages the "Nuclear Winter" concern well-publicized over a decade later, when scientists, including Carl Sagan, considered dire global impact of an all-out nuclear holocaust.[10] So a weather control device, not unlike the

Oxygen Destroyer that killed the original Godzilla, may be viewed as a weapon of mass destruction!

Scientists in *Son of Godzilla* are arrogantly endeavoring to transform huge swaths of our planet such as "useless" jungle, desert and tundra areas for the good of mankind into agricultural oases. Then we may grow crops to eliminate famine, feed the masses and reduce socio-political strife—oblivious of the fact that their actions, even if supported by the United Nations, would prove disastrous ecologically-speaking, as well as insensitive to certain, indigenous species (i.e., other than man).[11] The "diminishing food supply versus overpopulation" theme had been previously addressed onscreen in 1955's *Tarantula* (Universal-International, directed by Jack Arnold), although therein a mad scientist's solution was a synthetic, radioactive nutrient.

As Nobel laureate Sir MacFarlane Burnet stated in 1972, "Nature has always seemed to be working for a climax state, a provisionally stable system, reached by natural forces, and when we attempt to remold any such ecosystem, we must remember that Nature is working against us."[12] And espousing faith in technical innovation, Peter Vajk, author of *Doomsday Has Been Cancelled*, stated aggressively in 1978, "Should we find it desirable, we will be able to turn the Sahara desert into farms and forests, or remake the landscape of new England. ... We are the legitimate children of Gaia; we need not be ashamed that we are altering the landscapes and ecosystems of Earth."[13]

Rhoads and McCorkle add, the Solgell Island science fictional project and film "itself is prescient to several environmental concerns not only in late-1960s Japan but globally in the 21st century: overpopulation, desertification, destruction of the rainforests, and climate change."[14] Here is our classic "man vs. nature" theme projected into futurity. Yet, as Rhoads and McCorkle note, "humanity's attempts to control nature never end well ... the scientists instead open a nightmarish Pandora's Box."[15] Although not referring to Toho's science fictional films, former Vice President Al Gore noted that in the context of man's climate altering, global warming experiment (not unlike the fictional case with Solgell Island's scientists), "We are in fact conducting a massive, unprecedented—some say unethical—experiment. ... And ... many of the changes—such as the predicted extinction of half the living creatures on Earth—would be irreversible."[16]

But that mythical and metaphorical Box was already opened long before as evinced in Toho's flagship film pitting Godzilla against manmade pollution, *Godzilla vs. The Smog Monster* (aka *Godzilla vs. Hedorah*) directed by Yoshimitsu Banno. Rhoads dedicated his scholarly master's thesis to a study of this movie, laden with eco-conscious commentary, yet also a film usually derided as a chaotic, hallucinogenic, and silly if not confusing affair—quite a departure from Godzilla's prior on-screen exertions. Banno credited Rachel Carson's *Silent Spring* as inspirational for Godzilla's socio-politically relevant foe this go-around, stating, "I wanted to have (Godzilla) battle not with something like a giant lobster, but the most notorious thing in current society. ... What about a pollution monster?"[17] As Rhoads and McCorkle state, "*Godzilla vs. Hedorah* would prove to be a *kaiju eiga* protest film of a different order. ... a call-to-action ... regarding Japan's perceived transformation from an idyllic island nation to a woeful industrial wasteland."[18]

Hedorah is another one of Toho's continually and rapidly evolving-metamorphosing monsters that Godzilla battled (or became), like Mothra, Biollante, Orga and Battra, for example. Looming Hedorah begins as a tadpole-like creature. Quickly growing to gigantic size, perhaps in symbolic proportion to man's accelerative degradation of the environment, Hedorah attacks an oil tanker ship. Marine biologist Dr. Yano surmises that this strange mineralized life form represents a living form of pollution—an unwitting Frankensteinic resultant, a manufacturing byproduct created in the open environment inadvertently from our floating, accumulating wastes. This new monster's name is derived from the Japanese word for sludge—*hedoro*.

Attacks continue resulting in Hedorah's morphing into quadrupedal and then flying-saucer-like mutations—spewing toxicity and fumes wherever it moves. Eventually unnatural/anthropogenic Hedorah congeals, growing as it continually feeds off omnipresent industrial pollution, transforming further into a bipedal kaiju capable of even fending off Godzilla—now stricken with a vendetta to defend Japan against this disgusting, filthy yet deadly dooming horror, a creature spawned from the populace's own culpability via mismanagement of manufacturing waste materials and contaminants. Visuals of corpses resulting in the wake of Hedorah's attacks associates "threat of industrial pollution with wartime horrors and postwar nuclear fears."[19] Ultimately, Hedorah succumbs although audiences are warned of its possible resurgence from sludge, apprehensively capping this "bleak and poignant entry in *kaiju eiga* history."[20] Besides anti-pollution messaging the film criticizes Japanese counterculture of the time.[21]

Social commentary abounds in this film, as an economically revitalized post-war Japan was undergoing rapid awakening to their plight in consequence to "undeterred economic expansion."[22] As nuclear fears subsided during the 1960s (reflected in contemporary *kaiju eiga* films), unrestrained manufacturing with concomitant discharge of industrial wastes and unfiltered emissions mired Japan in a polluted hellhole.

Oil tanker spills fouled Japan's nearby ocean waters. As the populace became horrifically poisoned by such harmful occurrences and practices, environmentalism superseded nuclear angst as the primary socio-political and cultural concern.[23]

Japan's rivers and coastal waters, "literally the lifeblood of Japan,"[24] became dumping reservoirs for "enormous amounts of industrial byproducts." Manufacturing plants "spewed smog into (the) air. This toxic environmental degradation of Japan's land, water, and air became the focus of *kaiju eiga* productions in the early 1970s."[25] Fisheries, especially, suffered mightily as did afflicted humans from well-publicized incidences of methylmercury, formed as a result of mercury released to seabed marine sediments, followed by a process known as bioaccumulation—contaminant concentrations increasing organically up the food chain.[26] A 1972 report stated that Tokyo Bay, for one, "was devoid of marine life," representing a possible science fictional aftermath of Dr. Serizawa's Oxygen Destroyer weapon.[27] Furthermore, "Landfill reclamation projects also crowded the coastline, altering ... (marine) ecology while providing more room for increased industrial expansion and yet further pollution. ... red tides wiped out wild fish and marine creatures."[28]

One positive outcome was inauguration of Japan's Environment Agency in July 1971—coming a year after Richard Nixon's installation of the U.S. Environmental Protection Agency (inaugurated as an Agency on December 2, 1970). By 1972, public opinion in Japan had swayed considerably—now favoring remediation of their vast, self-imposed pollution problem, *even* if it meant industrial operations and economic expansion would be diminished. In particular, for our purposes, "The Godzilla and Gamera franchises of the early 1970s both emphasized the desperate state of Japan's environment, especially its seas and sea life. ... these *kaiju eiga* reflected a new nation-wide dialogue on Japan's environment-be-damned economic growth."[29]

Meanwhile another Japanese movie production company, Daiei, had been making movies since 1965 featuring its parallel kaiju to Godzilla, the gigantic, prehistoric chelonian Gamera, a series more commonly regarded as kiddie fare. In July 1971, with Hedorah the Smog Monster newly at large in Japanese cinemas, Gamera's then latest and seventh foe, Zigra, also swept over Japan. Author John LeMay considers this film to be "the first giant monster movie to emphasize the dangers of environmental pollution over that of nuclear testing. ... The film (is) ... rife with heavy-handed ecological messages."[30] favoring stewardship of the environment.

Although generally considered in America as one of the least satisfying of the early Gamera films, Rhoads and McCorkle recognize this rarity of sci-fi films *Gamera vs. Zigra* (directed by Yuasa Noriaki) for its overt ecological messaging. With many scenes shot at the then about-to-open Kamogawa Sea World, thematically this movie fortifies how man must protect oceans and its fragile marine life from our unceasing polluting tendencies. Underscoring this central idea, one marine biologist character, Dr. Ishikawa (Saeki Isamu), states, "These waters will become polluted. ... Progress in science destroys nature. We must protect animal life from pollution."

Titular Zigra is another planet, 480 light-years distant, which via spaceship has sent a representative "Woman X," whose human body—geologist "Sugawara" (both characters played by Yanami Eiko) from the Lunar Base—was taken over by the aliens to subjugate Earth through terrible cataclysms—earthquakes. She foreshadows our future if our polluting ways continue. Emphasizing the marine dilemma we now face, Woman x divulges, "We on the planet Zigra lived in the sea, but cultural progress polluted our waters and we could no longer live there. ... your seas are becoming polluted by sludge. If we leave Earth to you, sludge will cover your beautiful seas."

But Zigra is also the name of a huge monster (transported from Zigra) that can only be defeated by Gamera if Earth and therefore its bounteous seas are to survive. It resembles a weird swordfish/shark. It flies but can also walk bipedally on land; it emits powerful rays from its maw. Soon Zigra goes on a rampage, attacking an oil tanker. When Gamera dispatches the alien spaceship, Zigra grows to gigantic size because our water pressure is lower than on Zigra. Monster Zigra will devour fish in the sea, creating a shortage for humans, but then will also annihilate life on Earth. We are undeserving of our beautiful oceans. In the English-dubbed DVD version (with inconsistent subtitling) released by Shout! Factory, Zigrans will thrive in the sea yet (also) eat things living on the land. "In short: without the sea, Japan starves."[31] The not-so-subtle moral of the story summed at the end is that water is sacred and that "if oceans die, mankind will too." Or, as Oceanographer Patricia Tester later opined, "The oceans have become nothing but giant cesspools ... and you know what happens when you heat up a cesspool."[32]

By the late 1980s, with Hedorah far behind in the rearview mirror, Toho had largely moved on from the formerly prominent radiation theme (becoming gradually less central to latter daikaiju films of the 1960s) debuted in 1954's *Gojira*. Instead, ever increasingly throughout the following decade, Toho crafted stories around threats to humanity from bioengineering experiments gone awry (prior to 1990's *Jurassic*

Figure 25: Poster art advertising 1971's environmentally-themed *Gamera vs. Zigra* (author's collection, with appreciation to J. D. Lees).

Park novel), or those reflecting ecological and environmental "green" peril with concomitant human-induced potential for climate change. This is also the analogous trajectory undertaken by Legendary during the 2010s with its series of dino-monster films. Accordingly, next we shall briefly examine two of Toho's entries, *Godzilla vs. Biollante* (1989, but re-released on U.S. home video in 1992) and *Godzilla vs. Mothra* (1992).

In America we learned through Michael Crichton's 1990 novel *Jurassic Park* that man's recent attempts to harness genetic power will be potentially more harmful and dangerous than in the case of our half-century-old abuse of nuclear power. But a year earlier Toho had produced *Godzilla vs. Biollante* (directed by Omori Kazuki) in which a grieving Frankensteinian biologist, Dr. Shiragami (Koji Takahashi), preserves his deceased daughter's (Erika's) DNA encoded within a genetically-altered rose flower that subsequently becomes monstrously out of control and can only be defeated by Godzilla. Omori's incentive was born out of prescient forecasting that "a new breed of youthful human beings will appear, through genetic engineering."[33] If controlling climate (as in *Son of Godzilla*) isn't the proper means for feeding starving populations of the planet, then perhaps this circumstance may be resolved via biotechnology—man's creation of genetically "improved" crops.

Cellular tissue scraped from Godzilla's body during his destructive 1984 Tokyo rampage becomes unscrupulously manipulated. Central to *Godzilla vs. Biollante*'s plot lies the disturbing notion that tampering with natural genetic material result in artificial, or "unnatural" organisms and manmade ecosystems that will deleteriously alter our environment globally—overweening actions worthy of Gaian or Medean wrath. These cells may be grafted into cacti and wheat genes, creating a "super-plant" that will productively thrive agriculturally in desert regions of the planet. In *Son of Godzilla*, tragically, "scientists disregard the importance of ecosystems that do not directly benefit the human population."[34]

Of course, this genetic fusion process doesn't go well, as Erika's DNA and spirit are soon further incorporated into the chimeric hybrid rose organism containing Godzilla's cells. Such a circumstance is "totally different from what God intended for Earth." As the "tribrid" monster grows, its tendrils stirring from within, an "unsettling (musical) score indicates that, for the time being, biotechnology has surpassed nuclear energy as the Godzilla franchise's preeminent concern."[35] Biollante is in the broad context of giant *kaiju* monsterology, the conceptual cousin to *Jurassic World*'s Indominus rex—something new to the natural world, merging multiple species. In the end, Dr. Shiragami utters prophetically, "Godzilla and Biollante aren't monsters—the real monsters are those that created them."

But Godzilla's cells are also being used to create a defensive weapon in case of a nuclear accident, or in order to "neutralize nuclear missiles" … or even a nuclear-charged Godzilla! Yes, scientists are harvesting anti-nuclear energy bacteria from genetic material within Godzilla's tissues. This is their presumed, overt good intention, while Dr. Shiragami's infusion of Erika's spirit (genetically) into Biollante is the covert unwise, misguided, or bad intention. A theme here is, as stated in the movie, "If *we* don't do it, someone else will." Perhaps like China's recent pursuit of genetically enhanced super soldiers?

The outcome is rather analogous to the plot of the 1931 Universal *Frankenstein* movie where Dr. Frankenstein's assistant, Fritz, drops a container holding the good brain—representing the truncated culmination of a mad scientist's efforts to play god by creating a superior human being—then instead tragically grabbing the diseased brain of a disturbed individual that eventually becomes surgically inserted into the monster's cranium, thus dooming the lofty ideal of the mad scientist's experiment. Except in *Biollante* strands of both good (Erika's) and bad (Godzilla's) genetic material is intertwined, setting the stage for a very odd *23andme* type of family reunion on the battlefield (where a monstrously transformed Biollante is dispatched, allowing Erika's tortured soul to rest).

In *Godzilla vs. Mothra* (1992), directed by Takao Okawara, Earth must defend itself once more from the human scourge, as well as our most symbolic anthropogenic fiend—Godzilla, child of the atom. From beginning to end, this film is rife with eco-messaging, even including a Gaian-type example of a negative feedback loop involving Mothra's re-emergence that is central to the plot. No entry in Toho's filmic canon comes as close to Peter Ward's (later formulated) erudition of the Gaia vs. Medea concept.

While Akira Takarada's character, Joji Minamino, early on notes changes in global climate, sea level rise, increasingly powerful typhoons, and deforestation leading to enormous landslides, a colleague states, "I'm afraid we may be headed for Doomsday." Then to make matters much worse, Godzilla awakens in a deep-sea trench. When several explorers reach fabled Infant Island, our hidden archaeological past is revealed by a pair of diminutive, female "Cosmos" (Keiko Imamura and Sayaka Osawa), responsible for restoring Earth's natural order and maintaining balance (especially Japan's). Without invoking Gaia's name, the Cosmos comprehend that "Earth is a living being." Deforestation and an ensuing landslide on Infant Island has, in fact, disinterred Mothra's huge egg, suggesting a timely (Gaian-ish) negative feedback in play triggered at a crisis period when the environment deteriorates to such a degree—thus unleashing Mothra, Earth's guardian, to restore harmony.

It seems that Mothra had a not-too-long-ago conflict with humans and another gigantic insectoid creature known as the Black Mothra, or Battra. Twelve thousand years ago, a prior advanced human civilization developed a climate-controlling device which "offended Earth." Conflict ensued as Mothra defeated Battra, who was intent on destroying mankind. So while both Mothra and Battra are creatures innately tasked with environmental protection and restoration of Earth's natural harmony, only Mothra would prioritize preservation of human life in a geological crisis. One character in the film clarifies, "Mothra is *different* from Battra: it is trying to prevent disaster."

Battra remains unconcerned with mankind though and would instinctively seek to destroy us if we stray by harming the Earth. Thus, vengeful Battra anticipates an aspect of Peter Ward's (vengeful) destructive Medea ideology (or an "inured Gaia"), while Mothra together with the Cosmos may be viewed as a soft Gaian team. There is a suggestion that Earth's natural system and long-term, prolonged human civilization are mutually exclusive when ecological circumstances drift out of balance, in face of mass extinctions for example.

A larval form of Battra—manifestation of Earth's vengeance—awakens following an oceanic meteorite impact. This metaphorical meteorite, according to Mr. Minamino, is a "detonator" triggering a "time-bomb set by us humans." Apparently, because an undersea fault becomes destabilized and later Mount Fuji erupts! Someone declares, "Earth is getting its revenge."

Especially during the 1980s and early 1990s, geologists were finding numerous, then highly publicized connections between mass extinctions of the fossil record and large asteroid or comet impacts. In *Godzilla vs. Mothra* a world climate pattern develops that is highly analogous to how paleontologists view the dinosaurs' extinction. Paleontologists now have evidence that before the six-mile-diameter asteroid that impacted Earth 66 million years ago, extinguishing non-avian dinosaurs and other organisms, there was a prolonged period of global temperature increase due to massive carbon dioxide-emitting Late Cretaceous eruptions of the Deccan Traps volcanic region—resulting in ocean-wide acidification. Likewise, in *Godzilla vs. Mothra*, our global warming is noted as rapidly increasing; meanwhile in outer space a potentially mass extinctions–causing, Era-ending asteroid hurtles on trajectory toward Earth. Man's fate could thus symbolically follow that of the exterminated dinosaurs.

Firing sizzling rays from its eyes, Battra has an instinctive enmity toward Godzilla because the latter is a tainted by-product of misguided 20th-century human machination. And yet man must favor Mothra because Battra would destroy us (i.e., as would King Ghidorah). Later, adult forms of both Mothra and Battra team up to battle Godzilla, with Mothra assimilating Battra's spiritual energy—a weird fusion—upon its death, but not before they dispatch Godzilla. Then Battra communicates to Mothra that the large planet-devastating asteroid is on its way toward Earth, destined to impact in 1999. Mothra's promise to the dying, battered Battra is to deflect this object from its current course of eventual destruction, thus sparing Earth. The Cosmos warn as Mothra departs, "If the world lives to see another century, please remember what Mothra did for you and the planet you live on!" The moral is crystal clear ... we must learn to take good care of the world *ourselves*.

Several times giant monster movies have precipitated directly from *acute*, history-demarcating events. Cases in point—man's first nuclear war—*The Lucky Dragon* incident of 1954 (plus two atomic bombings in 1945) led to Toho's *Gojira*. Then America's 9–11 catastrophe became metaphorically captured on found footage via 2008's *Cloverfield*. That Man is ultimately the monster becomes vicariously concealed and substituted on-screen through visage of titanic invading creatures. And 2016's *Shin Godzilla* (Toho, directed by Hideaki Anno and Shinji Higuchi) was a visceral reaction to the Fukushima Daichii Nuclear Power Plant tidal wave–causing earthquake disaster of 2011. This movie rebooted the Godzilla series, harkening back to original fears of radiation exposure—once more emphasizing Japan's fate, with mankind therefore facing potentially planetary-wide adverse consequences to this insidious form of pollution. There is no referential (or reverential) suggestion of Gaia or Medea in *Shin Godzilla* however, thus distinguishing it from the Legendary series then unfolding.

In the 2016 film Godzilla first arises from the sea as a quasi-aquatic form, then rapidly mutates and evolves into a terrestrial biped as it traverses toward Tokyo. Its origin is stated as a result of oceanic radioactive waste dumping—materials which the developing Godzillean creature (originally an "ancient species of marine life") fed upon. Thus, this new Godzilla is closely akin to Hedorah and other filmic rad-waste pollution mutant monsters spawned in the sea. And yet, following four mutations, invulnerable, radiation-spewing Godzilla is regarded therein as "the "most evolved creature on the planet ... surpassing man in every way ... a god incarnate." The fear that this unknown Godzilla may self-propagate, or even that it or its progeny might conceivably sprout wings (who could refute then?) in a forthcoming mutation phase and spread to other countries and continents seems not only possible but palpable. Now it seems that only a one-megaton thermonuclear bomb deployed by the Americans might stop Godzilla in its tracks, despite sacrificing Tokyo, before things go too far.

But as in many Toho productions, Japanese scientists ingeniously devise a chemical weapon capable of halting and often defeating the giant monster menace. In this case, weaponized technology is founded upon a theoretical notion that Godzilla's high-temperature physiology is fueled by ingested radioactive elements. Therefore, radiation and occasional discharges from its core could be frozen using a blood coagulant injected into its body. The formulated coagulant solution is delivered via Operation Yashiori and Godzilla whose chest temperature has plummeted to minus-196 degrees F, is immobilized, that is, temporarily until the movie sequel comes out several years later (at which time, a nuke might be needed after all).

Shin Godzilla's possible origins as a "relic dinosaur" are discredited by scientists. In this case the rapid-driving evolutionary mechanism is man's nuclear pollution of the seas. While we may learn the true nature of those "reptiloid" osteoderm-shaped figures forming on its tail in a movie sequel, it's interesting how their vaguely reminiscent human shapes suggest an awful genetic tie between the monster and ourselves.

Contemporary sci-fi and horror movies aside, public environmental concerns had already escalated, prior to the Smog Monster's emergence in Tokyo Bay, that is, in the western world—perhaps most prominently and significantly in America decades earlier!

A Late 20th-Century American Environmental Movement

Although historians of science might consider Charles Darwin (1809–1882) the first modern ecologist, arguably America's environmental movement was ushered in with the publication of Rachel Carson's (1907–1964) bestselling book *Silent Spring* (1962). As she asserted, a normal pace of evolution simply cannot keep up with rapid changes inflicted upon Earth by mankind. For by 1942, man's "DDT era" had commenced.[36] For, like microbes (disease-carrying bacteria, viruses and "superbugs"), insects are imbued with a tantalizing genetic capacity to rapidly develop resistance to (once formerly) lethal chemicals, thus rendering these substances useless. Thus, the DDT era nearly coincided with the beginning of man's understanding of the "Age of Resistance."[37] By 1951, man's chemical attacks were regularly being thwarted as adaptive resistance developed within surviving (i.e., the most Darwinian fit) insect populations around the globe.[38]

Carson alerted readers to the dire problem of using chemicals to spray crops and insect pests, using prose descriptive of nuclear bomb ash fallout ... (reminiscent of what occurred, say, during *The Lucky Dragon* incident). "In the gutters under the eaves and between shingles of the roofs, a *white granular powder* still showed a few patches; some weeks before it had fallen like snow upon the roofs and lawns, the fields and streams. No witchcraft, no enemy action had silenced the rebirth of new life in this stricken world. The people had done it themselves."[39] Carson even acknowledged *The Lucky Dragon*'s human casualty, Aikichi Kuboyama (1914–1954), placing his fate directly in relation to that of a Swedish farmer who died from lethal exposure to sprayed insecticide powder. "For each man a poison drifting out of the sky carried a death sentence. For one it was radiation-poisoned ash; for the other, chemical dust."[40] But although this new deadly class of pollutants created for the war against bugs—insecticides—weren't radioactive, they proved every bit as harmful to humans and other vertebrates even at industrially prescribed dosages.

Referring to insecticides generally as the "partners of radiation," given that such mutagens and carcinogens may produce "precisely same effects,"[41] Carson astutely noted, "Along with the possibility of the extinction of mankind by nuclear war, the central problem of our age has therefore become the contamination of man's total environment with such substances of incredible potential for harm—substances that accumulate in the tissues of plants and animals and even penetrate the germ cells to shatter or alter the very material of heredity upon which the shape of the future depends."[42] Clearly, to Carson, considering effects on human health, the "parallel between chemicals and radiation is exact and inescapable."[43]

There *are* alternatives to wantonly spraying bugs with chemicals—substances that insects quickly re-adapt to, anyway—harmful to the environment; experimental repellant ultrasound techniques for instance, for deterring pestilent insects, as mentioned by Carson in 1962.[44] Here is an odd association between *Godzilla: King of the Monsters'* bioscientist Emma Russell and reality. Except in Legendary's science fictional mode, such a technique was used for "deep ecology" *evil* rather than for good of mankind ... and Earth. Another option would be to rely on bioengineering genetics to increase natural resistance in crops such that there would be reduced reliance on spraying pesticides and herbicides. (Shades of Biollante!)

Former Vice President and then Senator Al Gore addressed the pressing global environmental panic button in his well-reasoned, thoughtful, visionary 1992 book *Earth in the Balance: Ecology and the Human Spirit*, particularly stressing worrisome economic, social and political perspectives. But to those who practice "industrial alchemy,"[45] he also offered a bold plan leading toward resolution and redemption. (*Saturday Night Live*'s mock commercial for the "Yard-a-pult" offers no viable solution; neither is the "not-in-my-backyard" (NIMBY) syndrome broadly helpful.[46]) As Gore opined, "We should begin with the debate over global warming, because while it is only one of several strategic threats, it has become a powerful symbol of the larger crisis and a focus for the public debate about whether there really is a crisis at all."[47] A crux of the matter is that (generally) mankind no longer believes itself to be part of a *natural*, global system, instead supposing that we are somehow separate from Earth's life cycles[48] (my italics).

Yet the non-linearity of climate warming on Earth systems could only pose dangerous, catastrophic and unethical impacts on ecosystems, the likes of which we can at present only barely conceive.[49] Gore stated, "In the course of a single generation, we are in danger of changing the makeup of the global atmosphere far more dramatically than did any volcano in history, and the effects may persist for centuries to come. ... In the lifetimes of people now living, we may experience a 'year without a winter.'"[50] Exponential overpopulation of the planet with concomitant, rapid increases in (non-renewable) resource consumption, including the means to feed ourselves, remain at the root.[51] Relying on the development of higher-yield genetic crop varieties isn't a long-term solution.[52]

Foreshadowing bioscientist Emma Russell's twisted logic in 2019's *Godzilla: King of the Monsters*, Gore even suggested that

> the Deep Ecology (movement) considers human civilization a kind of planetary HIV virus, giving the Earth a "Gaian" form of AIDs, rendering it incapable of maintaining its resistance and immunity to our many insults to its health and equilibrium. Global warming is, in this metaphor, the fever that accompanies a victim's desperate effort to fight the invading virus whose waste products have begun to contaminate the normal metabolic processes of its host organism. ... And the internal logic of the metaphor points toward only one possible cure: eliminate people from the face of the Earth.[53]

(However, Gore firmly denounced such an endgame solution.)

Recognizing that our present situation isn't sustainable, Gore outlined what he referred to as his Global Marshall Plan. Founded on the premise of General George Marshall's recovery plan, successfully led by the U.S. in the aftermath of World War II to help restore the utter devastation of Europe, Gore's version of such a comprehensive effort to salvage the environment from its rapid decline would invoke a far more globalized view. Interestingly, the first strategic goal of a proposed environmental Global Marshall Plan would have been the weighty matter of stabilizing the world's human population. Another goal would involve properly cost-assessing—through formulated economic indicators—*and* discounting use, devaluation and depreciation of any plundered natural resources such as forests and eroded agricultural soil and polluted water bodies, as they are used to create products for sale.[54]

Some writers referred to the 20th-century's waning years as the Age of Ecology.[55] Certainly, "then" corresponds to America's historical period when environmental groups, organizations and activists became prominent and newsworthy, saving the environment. But to many beyond the early 1980s, it may have seemed as if nurturing Earth Day, first celebrated on April 22, 1970,[56] had instead spawned ranks of dangerous, biocentric eco-extremists! One such activist, Dave Foreman (an Earth First! cofounder), foreshadowed eco-terrorists' messaging in *Godzilla: King of the Monsters* (2019), stating, "It's time for a warrior society to rise up out of the Earth and throw itself in front of the juggernaut of destruction, to be antibodies against the human pox that's ravaging this precious beautiful planet."[57] Yes, a fever-causing *pox* ... the planetary infection caused by James Lovelock's "three deadly C's" operating in out-of-control fashion—combustion, cattle and chainsaws![58] In the eyes of author Christopher Manes, as well as those practicing civil disobedience and/or "ecotage"—enactment of illegal deeds conforming to "radical environmentalism"—the *fourth* deadly "C" would be civilization itself.

Radical environmentalists favor a return to the Garden of Eden, "where humanity lived in bliss."[59] This ideology doesn't sound too far removed from Alan Jonah's aspiration in *Godzilla: King of the Monsters*, in which eco-terrorists long to return to a golden, (Atlantean) classical age, sort of a New Age that's really old—recorded on ancient stone carvings, preserved as archaeological relics—prior to modernity when giant monsters like Godzilla lived in harmony with man. For if we don't somehow return to said garden, then mass extinctions might inexorably sweep across the planet, causing evolution itself to become forestalled. Civilization would cease in the wake of the Promethean "industrial monster."[60] And why? In part because ultimately "nature is not only more complex than we think, but it is more complex than we can ever think."[61]

In the eyes of deep ecologists, Nature is an entity we cannot fix, manipulate or control without Earth suffering vast consequences for such arrogance. So perhaps, as some cynically opined, it would be better during a pending apocalyptical coming third world war, if—in lieu of launching nukes—specialized biological weapons were invented for use that only harmed humans, while sparing (the rest of) the biosphere.[62] For instance, humans might not fare so well in our *next* pandemic war with lethal microbes.

Instability—as opposed to Gaian stability—reigned throughout the late 20th century. In 1994, underscoring the inescapable interrelationships of our planetary-wide woes, Laurie Garrett noted, "By the 1990s it was already obvious that the countries that were experiencing the most radical population growths were also those confronting the most rapid environmental degradations and worst scales of human suffering."[63] Her magisterial, in-depth analysis incorporated infestations of diseases spawned in part by global warming, stressed and fragile ecosystems (such as within marine settings increasingly plagued with toxic microbes), and rapid loss of biodiversity—especially within the vital Amazonian rain forest. Rapid population growth with concomitant added industrial and human waste pollution diminished susceptible ecosystems globally. Thus, many environmental settings were impacted, while further facilitating the transmission of harmful superbugs to innumerable hosts—retaining a capacity for human infection—throughout the biosphere.

Contrary to mid–20th-century ideals when scientists believed many or most diseases could be eradicated (like smallpox), now instead rapidly mutating, plasmid and gene-exchanging microorganisms—snippets of viral RNA—were winning on many fronts throughout our "global village." But were they succeeding virally—vengefully, in a Medean sense—to knock man off his pedestal? Was man finally being checked and countered by lethal, antibiotic-resistant disease outbreaks as part of an insidious (yet natural) global negative feedback system ... to restore *balance*?[64] Through the ages, some would assign such occurrences—plagues, pandemics—to (deliberate) godlike vengeance, whether presided over by God, a metaphorical God(-*zilla*), or even a mysterious Celtic or New Age Gaian/Medean deity.[65]

By the late 1980s, expert virologists reflected on such mutations, asking "What is the likelihood that a truly new virus capable of causing human disease will emerge?"[66] Or, why haven't viruses "wiped out all life on Earth?"[67] Furthermore, it seemed that by 1990, global warming would only sustain the worldwide microbial advance: "They are our predators ... evolving far more rapidly than *Homo sapiens*, adapting to changes in their environments by mutating, undergoing high-speed natural selection."[68] Later, the concern was over "any crackpot with a few thousand dollars' worth of equipment and a college biology education ... (who) could manufacture bugs that would make Ebola look like a walk around the park."[69]

Laurie Garrett further noted, "It wasn't necessary ... for Earth to undergo a 1- to 5-degree temperature shift in order for diseases to emerge. As events since 1960s had demonstrated, other, quite contemporary factors were at play. The ecological relationship between *Homo sapiens* and microbes had been out of balance for a long time."[70] Although unpredictable, yet so fearfully imaginable, what could be the worst possible scenario? Horror novel writers from Mary Shelley to George R. Stewart, or from Stephen King to Robert Kirkman seized the possibilities.[71] In one recent example, we read this relevant revelation: "What do viruses do? ... They multiply, adapt and kill. People did all that first, didn't we? ... We were the disease. ... What does that make the zombies? ... They're the antivirus. ... Because we let our virus run wild, the world had no choice but to put up a defense."[72] Meanwhile, postmodern producers of *dino-monster* films generally abandoned this fertile territory—once and formerly explored in 1953's *The Beast from 20,000 Fathoms.*[73]

Meanwhile, following his revelation half a century earlier, James Lovelock continued to write about Gaia on into the 2000s: one such title was 2009's *The Vanishing Face of Gaia: A Final Warning.*[74] Although still cast in the role as Gaia's chief spokesman, by this period Lovelock had become ever more pessimistic about civilization's survivability. Groping for viable options, he advised, "Earth has not evolved solely for our benefit, and any changes we make to it are at our own risk. ... it is hubris to think that we know how to save the Earth: our planet looks after itself. All that we can do is try to save ourselves."[75]

As Lovelock and others have pondered, it may simply be that Earth's carrying capacity for over eight billion people is no longer sustainable in the long run, especially given that most would prefer a highly greenhouse gas-generating, chemistry-for-better-living, westernized society and culture with all those industrial "perks" of consumerism! Continual use of oil, gas and coal might cause us to "become our own executioners and cause the death of Gaia. The Earth would then be left hot and barren with no life other than a few thermophiles: too sparse a biosphere for a self-regulating planet."[76]

Thinking cosmically, holistically and rather teleologically, Lovelock concedes,

> twentieth-century humans became almost a planetary disease organism. But it has taken Gaia 3.5 billion years to evolve an animal that can think and communicate its thoughts ... she has little chance of evolving another. ... As part of Gaia, our presence begins to make the planet sentient. We should be proud that we could be part of this huge step, one that may help Gaia survive as the sun continues its slow but ineluctable increase

of heat output, making the solar system an increasingly hostile future environment.[77]

Not quite "an evil religion" after all, rather, Darwinian "Gaia, like God helps those who help themselves. ... if we can evolve to become an integrated intelligence within Gaia, then together we could survive longer."[78] Of course, when forecasting events that only might happen, the unforeseeable always has potentiality. As *Jurassic Park*'s chaotician Ian Malcolm would comprehend, there are many chance or probabilistic paths leading toward the eventual possibility of any of several global apocalyptical changes—Ray Bradbury's "butterfly effect." And such changes may not trend in a more predictable, linear fashion.[79] But how can we facilitate Gaia's recovery? According to Lovelock, through a broader use of nuclear reactors!

Aside from the relatively slight danger of exposure to radiation from nuclear power stations, Lovelock believes there are practical reasons for more wholly relying on nuclear reactors instead of fossil fuels as chief energy sources. For one thing, Earth would not heat up as readily. And for another thing, the waste raw tonnage of using nuclear fuel per year is considerably less than carbon dioxide emitted through the burning of fossil fuels. But there is bitter irony in resorting to nuclear, as Lovelock outlines in his *Vanishing Face of Gaia*, published only two years prior to Japan's tragic earthquake-triggered Fukushima incident.[80] Lovelock believes that nuclear has been given a bad rap, but in light of our current plight it must be reconsidered as a viable energy alternative to other less practical means (the latter including some technologies that are generally considered green technologies, such as wind energy).

Lest we become another lost civilization, would our time on Earth become referred to as the "benighted Deforestation Age"?[81] Remember *Star Trek IV: The Voyage Home* (Paramount Pictures, directed by Leonard Nimoy, 1986), in which a mysterious interstellar probe approaches Earth in the year 2286, seeking the sound of live humpback whales (which in the plot have been rendered extinct centuries before)? Until their answering call responds, the cosmic probe wreaks havoc upon our planet—subjected to storms of biblical proportions, surely dooming our (futuristic) civilization.[82] Of course Captain Kirk and his stalwart crew resourcefully save the day, restoring whales to our future that signal the orbiting doomsday machine—a message that all is well here. But the message rings clear. Gaia requires a *diverse* ecosystem, and such plain need is absolutely universal. Indeed, scientists have determined that biodiversity is key to maintaining a proper oxygen level fit for the Anthropocene Era.[83]

While Covid-19 still ravaged during early 2021, even titans Kong and Godzilla became casualties, necessarily taking a back seat with a movie industry afflicted by the coronavirus pandemic—the "biological chain reaction," a pernicious (Medea-stirred) virus.[84] So as things transpired, *Godzilla vs. Kong* wasn't quite the nostalgic trip I'd been so long anticipating. (You can never go back, after all.) While the Americanized, original 1963 movie proved such an inspirational moment for me, everything about the aura of the new epic seemed ... off, fraught.

First, release dates kept shifting around. Then the new film finally opened during yet another surge in America's coronavirus pandemic. So, necessarily, did venues shift. While in my then half-vaccinated state (only one dose) I could have more riskily purchased a ticket for a real theater seat, I opted instead for a safer HBO Max home-screening. Such aspects, including the living room couch and a much smaller screen, diminished gravity of the moment. Or maybe I'm just too old to be enchanted the way I once was at nearly 9 years of age. Even so, it's as if Kong and Godzilla were forced to retreat in the face of that strange science fictional/apocalyptically resounding viral force. In the end, before their resounding rally cry, Earth's virus nearly defeated our monstrous saviors!

Whereas in *Pacific Rim* robotic Jaegers are humanity's saviors in a desperate battle to protect civilization as well as Earth's biosphere, in *Godzilla vs. Kong*, animatronic construct Mechagodzilla embodies our unnatural bioengineered directive—a mad-tech curse. To aspire to become an Apex species by exploiting any (Darwinian) means, in this case by defeating two natural varieties (Kong and Godzilla), perhaps is a human genetic inclination, or impulse. Kong and Godzilla symbolize a balance, harmony with the natural order that arguably cannot be restored as in classical antiquity, or one that is nearly as physically unattainable as reaching Earth's Edenic Hollow core. But via an unanticipated system failure, control over Medean Mechagodzilla is lost temporarily and the daunting Apex Cybernetics monster is dispatched before it nearly destroys Earth's two foremost Gaian operators.

As I received my second dose of Pfizer's COVID-19 vaccine, one day after *Godzilla vs. Kong* finally debuted on HBO Max, a twisted, stray thought entered my mind, reflecting on actor Raymond Burr (playing Steve Martin—not the comedian of latter-day fame) optimistically concluding 1956's *Godzilla, King of the Monsters* (the first "G-film" I ever saw, televised) with this remark: "The menace was gone ... the whole world could wake up and live again." Well, yes, hopefully and possibly, that is, until our next global crisis unfolds. In the context of geohistory, from a Gaian perspective perhaps, normalcy doesn't include mankind.

I've been asked before: do I believe in climate change? Long-term climate forecasting has its flaws, yet we suspect performed with some reliability, scientifically. Earth *is* warming and glacial ice is ever increasingly melting—as data shows. Such extreme weather patterns and social consequences persistently dealt with in recent years wouldn't otherwise have been anticipated. Anthropogenic changes may be superimposed over natural rhythms (such as exacerbated by super-volcanic eruptions of prehistory) and if destabilizing changes are geologically rapid, many species (including our own) will find it proportionally difficult to adjust (evolutionarily). And for us, given potential adversities for civilization—as we know it—isn't that the bottom line? And yet, on our ability to survive this dilemma I remain cautiously optimistic—but, alas, no better than that.

Tyrannosaurus rex (Latin for "Tyrant Lizard King") may be king of dinosaurs, but Godzilla—creatively derived in part from immortal *T. rex*—remains king of the monsters (as we know from a pair of movie titles). Kong—whose primeval ancestor ruled, wielding a bone-axe, osteoderm-bearing scepter, from a throne in a temple in an underworld Hollow Earth—somehow seems less monstrous to us puny, yet evolutionarily related apes. Meanwhile Gaia reigns supreme as Mother Earth Goddess.

Throughout the Legendary (with Warner Brothers) series of 2010s dino-derived giant monster films discussed here, note that the words "Gaia" and "Medea" are never spoken. However, the essence of Gaian principles is more than merely tangential. Goddess Gaia/Medea is implied or apparent throughout, albeit lurking as fringe pseudo-science. In the climactic film *Godzilla vs. Kong*, Hollow Earth speculation is more prominent than Gaian ecosystem theory—bounds of which are traversed more fully in earlier films such as 2019's *Godzilla: King of the Monsters*. It may be that Gaia theory, which had been railed against for many years by certain scientists for being not scientific enough, from the movie industry's perspective was ironically considered *too* scientific in nitty-gritty detail for public consumption and sloggy, within a fast-paced framework of entertaining, special effects laden giant monster movies. (In contrast, less strapped for running time, Greg Keyes elaborated more suggestively on possible Gaian eco-influences in his two series novelizations more so than audiences hear in scripted movie dialogue.)

We in our shared heritage with primate Kong have lingered in a primeval war with "dream dragons," symbolic Godzilla—condemning Nature—for far too long. Will the prophetic dinosaur ultimately win after all? Or should we now end the strife and finally, likewise as in *Godzilla vs. Kong*, move forward in alliance? While my fascination with Kong and Godzilla stems from fond childhood reminiscence, now as a senior citizen, their footsteps lead toward a stark projected reality via tenets of Gaian science. In the end, and despite their clans' vendetta, Kong realizes the broader scope his special role and presence encompasses—beyond Skull Island—while Godzilla has learned that he must share his alpha status with a former rival to maintain balance in the Anthropocene ... an epoch which, regardless, is destined to end someday.

Kong, Godzilla and Gaia have each been celebrated and worshipped as gods in fiction and popular culture by fans and cultists—*we* who have spawned a ravaging, gargantuan, militaristic industrial monster that may lead us all to apocalypse. It does not matter whether Mesozoic's (i.e., Godzilla) conquers Cenozoic's (i.e., Kong), or vice versa, during our sullen Anthropocene. But remember, there is no "Planet B" out in space for us to move to and live upon, at least not yet. So, until our giant (anachronistically prehistoric) metaphorical monsters either vengefully extinguish us *before* Earth is irrevocably devolved into a near-sterilized Precambrian state, or we learn to live in a harmonious New Dawn, Gaia/Medea shall remain true ultimate Ruler-Goddess of our fragile dwelling place.

But c'mon—get real. Godzilla and Kong are cool but fictional, only seen in movies, right? Meanwhile, complex, intricately interwoven Gaia essentially comprises the *reality* of our evolving Earth System. To whom (or what) should we and the monsters we so love bow? And what role shall Chaos or Providence play in our real-world predicament?

Can we rely on cooperative efforts of a filmic giant ape and a huge radioactive faux dinosaur to restore Earth's proper balance? Well, of course not in reality, but at the movies, audiences are mindful if not hopeful that projected metaphoric fantasy may transcend into reality. In *Godzilla vs. Kong*, "one will fall," and the vanquished ultimately will always be Mechagodzilla—hubris.

An iconic NASA photo of "Earth rise" taken by astronauts orbiting our Moon in December 1968 has become a modern symbol of the environmental movement. Can we slow the tempo of our global rapacity, thus stemming the tide of an avenging Gaia—all too ready to move on into a new era marked by man's absence? Can we thwart a seemingly inevitable, punctuated Medean consequence, this time in our Anthropocene, simply by becoming that *intelligent* species which learned to live within our planet's bountiful means?

Appendix

An Assortment of Titans, Dino-Daikaiju and Dino-Monsters

In order to save Earth and its myriad human inhabitants from utter annihilation, ultra-natural services and interactions of creatures vastly more powerful than mere dinosaurs and other prehistoric beasts that have survived the test of time on some science fictional lost island or ultra-planetary setting—no matter their natural great sizes or ferocities—may be required. Thus, we turn to the dino-daikaiju breed of monsters so revered by fans, film producers and writers of fantastic fiction, for salvation or abject condemnation. In this section a proper definition for a dino-daikaiju will be considered, followed by candidate creatures discussed in this book that may meet such criteria, or at least come close to the mark, having some qualifications. There are a number of interrelated terms that should be mentioned as well.

Monster movie fans in-the-know think they *know* what a kaiju is, right? "Kaiju" is a term in popular lexicon so overused it has nearly lost exactitude. While those enormous dino-derivative flesh monsters in *Pacific Rim* from the Anteverse known as *kaiju* ("strange or mysterious beast, or monster") unleashed a *G-Fan*-ish term into common parlance, the meaning of *daikaiju* is still relatively unknown, except to fans of Japanese monster movies, or those who read *G-Fan* magazine. However, in 2005, Robert Hood and Robert Pen, editors of *Daikaiju! Giant Monster Tales*, proffered a suitable working definition.

> As a term, *daikaiju* (Japanese, "*dai*" = giant + "*kaiju*" = monster) is usually applied to a distinctive form of fantasy film—a pseudo-SF sub-genre that began with Ishiro Honda's original *Gojira* in 1954 and was elaborated over subsequent decades, principally by Honda himself and his film studio Toho. ... To us daikaiju tales require monsters of unreasonable size, impossible and outlandish dimension, relativities that border

on (and sometimes cross into) the utterly absurd. Despite whatever ratio-nalizations might be applied within the narrative, daikaiju are fantasti-cal and provoke awe through the sheer audacity of their conception. ... but surely size is not all? ... A penchant for city-trashing and apocalyptic destruction, Metaphorical undercurrents. A sense that the kaiju are more than just Beasts—personality, in other words, albeit of a non-human kind. Pseudo-scientific and metaphysical pretensions. Vast scope. Incredible power. A certain cosmic inevitability. Daikaiju are not scared of Man. ... If they are often the unnatural product of human arrogance, they manage to transcend that heritage and become supra-natural. They are more like inhuman gods than unnatural beasts. They are impossible, yet they *are*.[1]

Author Jeremy Robinson has his own take on the subject of kaiju. In his astute estimation, kaiju species featured in his *Nemesis* novel series may be described accordingly: "Yeah, *Kaiju*. The word that came to define the giant monster genre that includes city-stompers like Godzilla and Gamera has become our official term for any creature that is ... well, not natural, with the understanding that it be reserved for things capa-ble of mass destruction. A snail with tentacles wouldn't qualify—unless it was ten stories tall."[2] Furthermore, a goal he hoped to achieve via his original Nemesis monster characters was establishing "America's first real iconic kaiju."[3]

J. D. Lees defined kaiju as creatures having a triad of characteris-tics. Briefly, in his first estimation, "Its size must beyond the upper limit of natural and enable the entity to destroy buildings. It must present a menacing aspect. There must be something inexplicable in its origin or nature." However, Lees decided that the third criterion was "superflu-ous" and so he replaced it: "It must show some degree of purposeful; intelligence beyond that of any comparable animal."[4] My opinion is that the "inexplicability of origin and nature" term could still be retained as a useful fourth "decider" criterion to settle certain cases. Lees also con-sidered the possibility of a more elaborate kaiju "classification" system, but this treatment has yet to transpire.[5] Must a set of kaiju criteria and qualities be numerically "weighted"? I'll leave such questions for greater minds to ponder.

In 2002 paleontologist Jose Luis Sanz outlined his views of Japa-nese *Kaiju Eiga* (e.g., "strange beast movie"), focusing on appearances featured therein, "typically combinations of various types of animals. Many of them are of prehistoric origin, although, as we shall see later, they may arise by a multitude of other factors, both of terrestrial and extraterrestrial origin. ... The *kaiju eiga* are closely associated with the oriental dragon."[6] Sanz further categorizes appearances of filmic dino-saurs into two general types: "real dinosaurs and dinosauroids. In turn the latter can be subdivided into three groups ... 'paradinosauroids,' 'sauriodinosauroids,' and 'dragodinosauroids.'"[7] Sanz's last term, dragodinosauroids, generally refers to those "characterized by being created by an actor within an ostentatious rubber suit. These are the formidable creatures of Japanese cinema, which have made possible the development of one of the richest and most complex strands of mythology within dinosaur cinema."[8] Godzilla would certainly qualify as Sanz's quintessential dragodinosauroid.

Dino-monsters are usually fantasy reptilian dinosauria of sci-fi literary or filmic persuasion that menace humans and are often of hugely scary sizes or fierce disposition, but what then is a dino-daikaiju (my invention)? Essentially and generally, these would be the (giant) forms of daikaiju appearing in kaiju eiga or other films, or as featured in printed stories and graphic novels, that resemble, mimic or would appear to be of oft-anachronistic prehistoric genetic stock—dinosaurians defying physics with super powers while congruent with Lees's general criteria. Some of these would be dragodinosauroids, while others might have other morphologies. Yes, they're impossible, but just *are*.

While Godzilla clearly is *the* ancestral dino-daikaiju of kaiju eiga, King Kong is an interesting example, if not the absolute test case. Kong, being a mammal, cannot be a *dino*-daikaiju. But is dinosaur-battling Kong a, well, ordinary kaiju? Ah, yes—finally that burning question, or dare I say the 100,000-pound gorilla in the room! Other than their sizes and appearance, persona forms part of the criterion. Some of us are infatuated with giant monsters; others prefer those that appear prehistoric but are still science fictional, and to many of these creatures the term "kaiju" is often loosely applied. In *G-Fan* no. 78 (Fall 2006, p. 72), J. D. Lees questioned whether King Kong is a kaiju. He opined definitively:

> The Toho Kong certainly qualifies, but the American version(s) are iffy, on two counts. First at 22 feet or so, (RKO's) Kong is large, but it doesn't require rewriting the laws of physics to imagine a mammal of such proportions. The prehistoric mammal "Baluchitherium" topped out at 25 feet. So Kong's size, though impressive, may not be beyond the upper limit of what could occur naturally. Secondly, Kong's creators have taken pains to retain animal-like behavior in the big ape, giving doubt to whether he possesses any human-level reasoning ability. Unarguably, Kong does exhibit some humanlike behaviors and expressions, but not necessarily any beyond what most lay-people would consider possible for a gorilla. If push came to shove. ... I would no longer include Kong as a kaiju, even though I have considered him one since my childhood ... in his American incarnation, Kong doesn't make the cut.[9]

Figure 26: Poster art for RKO's *King Kong* (1933). According to J. D. Lees, this was Kong in his least Kaiju performance (author's collection).

Of course, that determination depends on *which* Kong you're referring to.[10]

The new millennial Legendary movie series features Titans (Kong included) that are immense, legendary creatures, some formerly coexisting symbiotically with ancient man who regarded them as gods, recorded in history stretching back at least to the writing of "The Epic of Gilgamesh" (c. 2000 BCE). Titans might all qualify loosely as kaiju, although only certain species would be daikaiju (or even dino-monsters). As Keyes described them, Titans are regarded as part of the natural order, but

> didn't arise when the rest of life on Earth did. What if they weren't part of life as we know it at all? Maybe they came from before, when there no water or free oxygen, when everything was a volcanic hellscape, the atmosphere a perpetual lightning storm, when radiation sleeted from the sky and pulsed from the ground at levels that would strike a human dead in the time it took to draw a breath of the poisonous atmosphere. ... Before bacteria. ... Plenty of time for another kind of life to evolve based on some other chemistry. ... when the rains arrived ... only a few adapted, survived to live in an oxygen atmosphere, became immortal.[11]

Somehow, though, such a harsh description would seem to exclude mammalian Kong, who requires oxygen.

When reanimated from hibernation, Titans—known both from the fossil and archaeological records—have distinctive bio-acoustical characteristics and radiation signatures. Furthermore, during the Titan assault, triggered initially by Emma's ORCA device, and later as summoned by Monster Zero-Ghidorah, unleashing of an out-of-human-control giant monster wave causes "earthquakes, wildfires, tsunamis, and disasters we don't even have names for yet."[12] In the end, all surviving kaiju—sans Kong, who didn't show up in 2019's cataclysmic battle against Ghidorah—ultimately bow to Godzilla in the wake of Boston's destruction, "acknowledging their alpha ... as Serizawa's natural order was restored."[13]

Well now—is Lovelock's (and Ward's) Gaia/Medea herself a daikaiju, per definitions outlined above? After all, she certainly meets Lees's size criterion. In a well-balanced system, everything is precariously interconnected. Life and living on this planet is perpetually risky beyond comparison, per Darwinian law; although on Earth it's best to live within means. Despite centuries of theoretical intrigue, Earth's and life's intertwined early origins—her very nature—remain rather "inexplicable." And "purposeful intelligence"? Well, to preserve her longevity, in an eruptive fit of volcanic rage Gaia/Medea could exterminate many cities, falling like a line of dominoes, let alone species from her sphere, targeting those posing the most danger to her (less toxic or alarming) children—evolutionary experiments—or, during the Anthropocene, especially those who threaten global sterilization—us. The root of Gaia/ Medea's intelligence lies in her know-it-all capabilities as a long-term survivor—no matter what! Furthermore, like those majestic, primordial Titans viewed by the ancients as gods, Gaia has been considered a goddess—sometimes merciful while at other times merciless—in New Age circles.

So now let's briefly consider those giant, impossible monsters we so love to fear—daikaiju, Titans, dino-monsters and kaiju eiga one and all, including several of their non-kaiju yet eco-associated ilk—mentioned herein, or elsewhere in referenced literature. While this list emphasizes Legendary's biota (and then some beyond), note that it is not comprehensive to the genre at large. Perhaps such a fuller listing merits a complete book someday.

Agathaumas: One of the earliest horned dinosaurs, known to late 19th-century paleontology, this quadrupedal creature, whose visage was first disseminated in popular culture by Charles R. Knight through an 1897 painted restoration, had several starring stop-motion animated roles in movies of the early 20th century, such as in 1925's *The Lost World*. In part, Knight's painting lent Agathaumas's distinctive spiky head shield to that seen in Toho's Anguirus.[14]

Figure 27: Charles R. Knight's life restoration of the cryptic horned dinosaur, Agathaumas, known from scanty paleontological remains. A time-honored, usual suspect character of classic science fiction films, such as 1925's *The Lost World*, and early fantasy pop-cultural literature. This image was printed in *The Century Magazine* LV (November 1897).

Anguirus: This pseudo-ankylosaurian, part ceratopsian (e.g., horned) quadruped chimera was Godzilla's first prehistoric rival in Toho's 1955 production, *Godzilla Raids Again*. Although Legendary chose not to include this monster in its new millennial Monsterverse, Anguirus remains a lesser celebrated favorite among G-fans, especially for its role in the Americanized *Gigantis, the Fire Monster* (1959). However this dino-monster does evidently appear, battling Kong, in Colosi's script novelization as the "Evil Armadillosaurus." (See Chapter Nine.) Rather like Kong and Ghidorah in later filmic series roles and reincarnations, Anguirus also seems to have a prehistoric instinctive yearning, or Mesozoic predilection, to battle Godzilla—although now in modernity, the Anthropocene. After tumultuous city-fighting, the original Anguirus was destroyed by Gigantis/Toho's "2nd Godzilla" in the 1955/59 production. Possibly qualifies as a near Titan. (Also see entry for Margygr.)

Antarctic dino-daikaiju: These colossal dinosaurians—Frankensteinic abominations, originating via an alternate evolutionary process— terrorized intrepid explorers to the vast, icy southern continent in John Taines (i.e., Eric Temple Bell's) 1929 classic science fiction novel *The Greatest Adventure.*[15]

Ape Gigans: Mentioned in Chapter Four, the Ape Gigans is arguably the original "Kong-ish" monster-creature. Appearing in an 1872 English edition of Jules Verne's *Journey to the Center of the Earth*, the chapter in which it appeared was evidently never penned by Verne himself. It is intriguing that the anonymous author of this dream-like passage was inspired to incorporate a prehistoric ape-like creature, inserted within what is essentially a paleontological time-travel tale through Earth's geo-history, with intrepid explorers entering and leaving its planetary bowels via volcanoes. (I've always enjoyed this chapter of the novel anyway.) Ape Gigans's appearance is all too brief with

Figure 28: Jack Arata's 2005 illustration showing Ape Gigans battling the shark-crocodile in a 19th century English translation of Jules Verne's *A Journey to the Center of the Earth*. Original art made for hire (author's collection).

which to discern whether or to what extent it may have been kaiju. Another subconscious monster from the Id. See Chapter Four for more on this unlikely species.

Arachnida Malum (aka Mother Longlegs): A throwback to the (maybe not so completely) lost Spider Pit footage associated with RKO's *King Kong* 1933 classic, this gigantic 18- to 23-foot-tall spider is witnessed piercing a human victim on Skull Island in Legendary's 2017 movie *Kong: Skull Island*. There is further information on this species divulged in endnotes to Arvid Nelson's and Zid's graphic novel *Skull Island: The Birth of Kong* (December 2017). Here one reads that Mother Longlegs is a supermutation of those harvestmen colloquially known as Daddy Longlegs often encountered in suburban backyards. Also, "Mother Longlegs has evolved sharpened spikes at the end of its powerful legs. Eight razor-sharp blades rein down from above on unsuspecting prey, pinning them to the ground with a poisoned spur that immobilizes the victim and sucks nutrients from the body like a milkshake through a straw."[16]

Arctic Giant: Whereas sauropods were early favored cartoon-animated dino-monsters of film and comics, after *King Kong*'s release in 1933, theropods began making their mighty intrusion into the fray, beginning perhaps with a vicious, blood-thirsty tyrannosaur featured in Disney's "Rite of Spring" segment, directed by Paul Satterfield, from 1940s *Fantasia* (Disney/RKO). But a more terrifying and imposing 100-foot-tall pseudo-tyrannosaur, sporting a row of rather enlarged fins along its vertebral column, appeared in the 1942 Superman cartoon *The Arctic Giant* (directed by Dave Fleischer for Fleischer Studios). This pre-nuclear precursor to Godzilla was discovered frozen in suspended animation in Siberia. When it thaws, it goes on a building-smashing rampage through Metropolis.[17]

Battra: Akin to Mothra, and ultimately one of Earth's savior species, appearing in Toho's *Godzilla vs. Mothra*. See the Epilogue for more.

Brontosaurus: For decades, dino-aficionados "in the know" knew to properly refer to this dinosaur as *Apatosaurus*, although now both genera are recognized. By the late 19th century, long-necked quadruped *Brontosaurus* was perhaps the original dino-monster that was huge, extinct and sometimes could present a fierce disposition (as in 1925's *The Lost World*, or 1933's *King Kong*). In popular culture, however, North America's Late Cretaceous *Tyrannosaurus* eventually eclipsed *Brontosaurus* as most representative of the dinosaurian clan, especially in sci-fi film and literature. (The 1993 movie *Jurassic Park* didn't even feature *Brontosaurus*, once the most famous dinosaur known to have inhabited the Jurassic Period!) In a broad sense though, although not a daikaiju, the concept of a live (symbolic) prehistoric *Brontosaurus* (or its sauropod cousins then known to contemporary science) anachronistically surviving to our present or even invading our cities paved the way toward Godzilla's and Kong's reign on modernity.

Camazotz: This creature, named as such in Keyes's 2021 novel evidently is the bat-winged slithery-tailed winged monster that attacks Kong shortly after his arrival in Hollow Earth. However, as divulged by Keyes, one of these aberrations also emerged from Skull Island's underlying Vortex after the Ghidorah debacle, with Kong subduing this creature—as he had numerous times prior with resident Skull Crawlers. Camazotz is a fierce kaiju but arguably not a true Titan. Every giant creature that evolved in Hollow Earth perhaps does not merit that esteemed title. Only those which interacted with ancient humans rather in symbiotic fashion and have an overwhelming, overpowering presence on outer Surficial Earth are worthy of being named true Titans.

Cloverfield Monster: "Something has found us." (tagline) Towering kaiju-esque creatures of absurdly-looking yet powerful morphology invading Earth from another dimension, first appearing in 2008's *Cloverfield* (Paramount/Bad Robot, directed by Matt Reeves). The underlying science fictional concept is analogous to *Pacific Rim's* otherworldly Anteverse, although the Cloverfield monsters are not dino-daikaiju.

Daikaiju A: Appearing in D.G. Valdron's incredible short story "Fossils," published in *Daikaiju! Giant Monster Tales* (Agog Press, 2005). As I stated previously,

> Valdron's brooding tale of a Tokyo deserted by all except for "*daikaiju A*" and a few obsessed human observers suggests dinosaurian kaiju originated in the Mesozoic. In particular, "A," described as a radioactive god "impervious to time," sporting a glowing dorsal fin, is an allosaurian mutation—or possibly derived from plated stegosaurian, or finned spinosaurian stock. Or as Valdron professes, "The thirty-six known kaiju constitute a taxonomist nightmare. Most are clearly reptiles, relatively identifiable offshoots of archosaur lines such as dinosaurs. Their origins are placed between eighty- and two-hundred million years ago…" According to Valdron, in prehistoric times, natural radiation was stronger and more prevalent than today (as Lovelock would agree). Depositional processes concentrated radioactive sediments millions of years ago; dead animals (dinosaurs, etc.) were buried in these toxic environments. But as radioactivity acted upon DNA in tissues, eventually these corpses transformed, arising into indestructible, godzillean-sized *daikaiju*, non-living yet "walking nuclear reactors," the titular fossils of Valdron's story.[18]

Death Jackal: Although not animated for, or seen in Legendary's 2017 *Kong: Skull Island*, these 12-foot-long canine hyperenergetic carnivores haunt pages of Arvid Nelson's and Zid's graphic novel *Skull Island: The Birth of Kong* (December 2017). They are perhaps analogous (although descendant from far differing evolutionary progeny) with respect to the dino-monster Death Runners appearing in Joe DeVito's intriguing volume *Kong: King of Skull Island* (2004). (See entry for Gaw.) Death Jackals are not *kaiju eiga*.

Dimetrodon: The classic eight- to ten-foot-long "sailback" or fin-backed carnivorous quadrupedal pelycosaur from the Late Permian, and discovered by scientists in the late 19th century. This animal was more closely related to the base of our mammalian lineage than ancient reptilian stock. In the evolutionary context it was intermediate. However, it is often misconstrued as a dinosaur in the popular vein (which it is not, but in our dignified role as fans of sundry sci-fi and horror films we may loosely consider it to be yet another dino-monster). Monster movie/sci-fi fans, for example, will fondly recall those 50-foot-long dimetrodonts featured in 1959's *Journey to the Center of the Earth* (20th Century–Fox, directed by Henry Levin), on the shore of the "central sea"—an environment not looking much like Kong's Hollow Earth.[19]

Dino: This particular name reference is extended toward Winsor McCay's illustrated cartoon Dino, which would have appeared during the early/mid–1930s as a newspaper serial feature on the comics page. In draft panel sheets Dino appears gigantic, somewhat daunting in demeanor, and wherever he strides the brontosaur inadvertently causes swaths of destruction in a metropolitan area. The creature is awkwardly gentle, though—not truly menacing, that is, in a Godzillean, giant monster-on-the-loose manner. It can be classified as a dino-monster, but not daikaiju. For considerably more on Dino, see Ulrich Merkl's lavish (2015) *Dinomania* book, or Chapter Twenty-six in my *Prehistoric Monster Mash* (2019) volume.

Gaia: See Chapter Three in this book and the Epilogue. Perhaps unexpectedly, James Lovelock's vision of Gaia rather qualifies as daikaiju. (Meanwhile, Peter Ward's personification of Earth-as-Medea instead might also qualify.)

Gamera: The gigantic, 200-foot-tall chelonian (prehistoric turtle) featured in a series of highly popular Daiei Motion Picture Company films, usually catering to a younger audience demographic than Toho's (or Legendary's) Godzilla series. Clearly this is a daikaiju, although one catering beneficially to mankind, children especially.

Gargantosaurus: A strange melding of man with alien entity morphing into an intelligent, shape-shifting paradinosauroid that haunts pages of William Schoell's gripping 1988 novel *Saurian* (Book Margins, Inc.). Although not a filmic creation, a daikaiju in its own right.

Gaw: This much enlarged spike-backed Raptor dino-monster succumbs to Kong in DeVito's and Brad Strickland's (with John Michlig) 2004 lavishly illustrated novel *Kong: King of Skull Island.* Think of what the *Velociraptor* lineage might have evolved into had they not been exterminated by the Cretaceous-Paleogene asteroid impact event 66 million years ago. They might have become the attendant feathered Deathrunners or their pack leader, the indomitable Gaw. Long-armed Gaw is cunning and bloodthirsty, much larger and more powerful than an ordinary tyrannosaur. In general morphology, Legendary's Indominus Rex seems rather reminiscent of Gaw—a near dino-daikaiju.

Gigantis: Toho's second, Americanized Godzilla—the one not eradicated by Dr. Serizawa's oxygen destroyer in 1954's *Gojira*—and conqueror of Anguirus. Truly a dino-daikaiju, maybe more so than the first Godzilla. (See next entry.)

Gigantosaurus: For Colosi's colossal sci-fi Gigantosaurus, see Chapter Nine. For the Gigantosaurus that was eventually described scientifically as *Brachiosaurus* (further split into *Giraffititan*), see Chapter Eleven.

Godzilla: Mountainous Dragodinosauroid, daikaiju, dino-monster, apex dinosaurian predator and mythical Titan! Legendary's version and vision of Godzilla differs from the evolving suitmation and metaphorical persona originally projected for this creature in Toho's prior stream of G-films, looking far more robust. Godzilla remains the quintessential *kaiju eiga*. Although originally cast as a radioactive manifestation resurrected from the Jurassic age through man's unwise tampering with the atom in the development of nuclear bombs, and later as a savior of mankind in battling creatures such as the alien Ghidrah/Ghidorah and Hedorah, Legendary's Godzilla is at his core not our enemy and, rather, judged by the younger Dr. Serizawa to be a benevolent monster. Once peacefully existing alongside ancient classical civilizations, Godzilla instinctively and resolutely protects the planet against unruly creatures—especially rival Titans such as Ghidorah and the giant buglike MUTOs—who might otherwise cause Gaia to sway out of balance. Throughout Legendary's Monsterverse, Godzilla seems to have forged a deep-rooted alliance with man despite our wayward means, despite our efforts to destroy him. One wonders how long this accord will be tolerated. The brutish, bulldog-like (CGI-enhanced) appearance of Legendary's robust Godzilla differs appreciably from Toho's historically rugged yet more svelte-looking suitmation varieties. (And of course, Godzilla's size—now 355 feet tall in Legendary's series—and mass has increased considerably throughout his numerous filmic transformations.)

Figure 29: Vincent Lynch's fantasy illustration of the Gigantosaurus (later scientifically described as Brachiosaurus) parading along a New York City street. The art appeared in the November 28, 1914, *Scientific American*, illustrating an article, "The Largest Known Dinosaur" (pp. 443, 446–447), by paleontologist W. D. Matthew. Brachiosaurus has since been surpassed in size and girth, although today Tyrannosaurus reigns as king dinosaur.

Gorgo: Although in the 1961 film *Gorgo*, the Gorgo semi-aquatic prehistoric species may not appear as daikaiju in a "Lees" sense, it may seem arguably more so in Carson Bingham's 1960 novelization from the movie script. See Chapter Twelve in my *Dinosaurs Ever Evolving* book (McFarland Publishers, 2016) for considerably more on classic dino-monster Gorgo.

Gorosaurus: This is the resident dinosaurian haunting the jungles of Mondo Island in 1967's *King Kong Escapes*. (See Chapter Four.) The bounding Gorosaurus is Toho's stand-in against Kong, analogously for the tyrannosaur which Kong defeated in the 1933 RKO film.

Hellhawks: Flying reptilian vultures encountered within Kong's "throne room" in Inner

Earth in *Godzilla vs. Kong*. Perhaps conceived as an homage to Edgar Rice Burroughs's Mahars of *At the Earth's Core* fame. To Ilene Andrews, these creatures, roughly two to three times larger than human size, struck her as resembling "griffins designed in hell."[20]

Indominus Rex: A genetic experiment (created in a laboratory merging tyrannosaur and velociraptor genes, with other species) gone awry, appearing only in Legendary's *Jurassic World* movie. This dino-monster is an instinctively rabid hunter and killer. Indominus is also a near daikaiju, based on its much larger than ordinary carnivorous dinosaur size and intelligent craftiness. And yet in an end struggle, perhaps aided by its much longer and deadly forearms, Indominus nearly kills the prehistoric theme park's resident *Tyrannosaurus*. In a grand competition for prehistoric survival in a modern world, pseudo-natural, yet genetically revived creatures (*T. rex* and *Mosasaurus*) are able to defeat the artificial, far less natural chimeric, monster. See image 4.

Indoraptor: This is another disastrous genetic experiment conducted on the sly by laboratory scientists seeking to perfect a new dinosaurian lineage. Appearing in *Jurassic World II*, the slightly larger than human-sized Indoraptor is part Indominus blended with *Velociraptor* genes. See Chapter One for more on this species. It is not daikaiju due to its less imposing size, but it is a deadly dino-monster.

Jinshin Mushi (aka Dragon Beetle, Titanus Jinshin-Mushi, MUTO Prime): Another of Godzilla's giant bug-like, carapace-shielded foes. As its eggs incubate, larvae feed from nuclear-rich hemoglobin in the host Titan. Gestation periods for embryos endure for centuries before hatching occurs through the ribcages of its victims. Their breeding constitutes a "force to consume all life on Earth," their ranks only to be held in check by a Gaian entity (e.g., negative feedback control) such as a Godzilla.[21] (From "Superspecies Profile" published in Legendary's 2019 prequel graphic novel *Aftershock*. See Chapter Ten for more.) Emma has had prior success (i.e., five years before the events in Legendary's 2019 film unfold) in defeating this kaiju using the ORCA.

Kong: Both Toho's and Legendary's Kong are of considerably greater mass and size than RKO's original Kong. Legendary's Kong however, is far more imposing than Toho's original Kong featured in their classic 1962/63 film. (See Chapter Four.)

Kraken (aka Titanus Na Kika): An aquatic Titan dwelling within the Indian Ocean, per Keyes's 2019 novelization, Kraken seems like the oversized octopus of legend that would drag ships into the depths with its tentacles. Like many of the Titans, Kraken is radioactive, gigantic, intelligent and can even, to the consternation of Monarch scientists monitoring its physiology, fake its own death. It makes one recall *It Came from Beneath the Sea* (directed by Robert Gordon, Columbia Pictures, 1955), or even that slimy, giant octopus which Kong fought in 1963's *King*

Kong vs. Godzilla (and another—see the entry for Mire Squid—that Kong defeated in 2017's *Kong: Skull Island*).

Kronos: An enormous, uncannily demonic *Kronosaurus* whose species has survived in oceanic recesses since the Cretaceous-Paleogene extinction event of 66 million years ago. Max Hawthorne's pliosaur is the central dino-monster in his 2013 horror novel *Kronos Rising* (Far from the Tree Press).

Leatherback: In *Pacific Rim*, this enormous, 270-foot-long creature battles the Jaegers in a tumultuous scene staged in Hong Kong. This Category-4 kaiju (per a system ranking ferocity on an ascending scale from one to five) appears "simian" in gait, has a protective carapace over its back and shoulders, a fluorescent spine, multiple eyes and "a biological electromagnetic weapon" that can discharge an electromagnetic pulse (EMP) like emitted from a nuclear blast.[22] Given their peculiar evolutionary history, all daikaiju in *Pacific Rim* may be classified as derivative dino-monsters. There's also a cult surrounding the Kaiju, such that many believe Anteverse "kaiju are a message from God, or the universe, telling humanity that we have broken the planet so badly we no longer deserve to live on it."[23] See image 3.

Leviathan: In Keyes's 2019 novel this Titan is associated with the world-famous, alleged Loch Ness sea-monster.

Magma Turtle: Unrelated to Gamera, this is a 34- to 40-foot-long herbivorous tortoise native to Skull Island in the Legendary series of graphic novels. Its eggs are hatched within a volcano, absorbing heat from adjacent molten rock. Even its carapace is fused during growth stages through life, forged by accreting cooled magma onto its shell.

Margygr: A whale-sized, armored reptilian Titan found dead in the Arctic. From its description found in Keyes's 2019 novel, Margygr seems possibly morphologically related to Toho's Anguirus.[24]

Mechagodzilla: In the Legendary–Warner 2021 film, this is clearly a daikaiju, constructed along saurian lines, perhaps the most powerful of them all, sans Ghidorah, although Mechagodzilla actually becomes the essence of Ghidorah. Mechagodzilla therefore projects an intelligence beyond most other natural creatures. After psionically linking with Ren Serizawa's neural pathways—rather in the fashion of a *Pacific Rim* Jaeger—Mechagodzilla rejected human control, deviating instead toward Ghidorah's dominance, as telepathically and vitally preserved in its skull matter and DNA. As mentioned in Chapter Ten, Mechagodzilla takes on a far more variant, unfamiliar form in Toho's 2018 anime film.

Mire Squid (aka Giganntus Leviapus): This 90- to 110-foot-long "Megalopod" is actually an omnivorous squid/octopus hybrid. Like many species native to Skull Island's strange ecology, its evolutionary mutation seems to have been fostered by a poorly understood process known as "superthermal gigantism."

Mokele Mbembe: Besides the African "living brontosaur" of 1980s pop-cultural cryptozoological fame, the name seems first used in relation to a fictional dinosaur in Henry Francis's 1908 short story (see Chapter Eleven). Legendary assigned this name to a snake-coiled Titan sporting a horn in Keyes's 2019 novelization, where this creature lays in suspended animation in Jebel Barkal, Sudan.

Monster Zero (aka Ghidorah, aka Ghidrah): This is the hellish, utterly unnatural creature erupting from Emma's ORCA, her Pandora's Box threatening the extinction of man. In Legendary's scope, Ghidorah is unnatural presumably because it is of alien, extraterrestrial origin. Long, serpentine-necked, three-headed, massively winged Ghidorah is invulnerable to an oxygen destroyer bomb deployed in *Godzilla: King of the Monsters*—unlike Godzilla, who nearly succumbs to this weapon in the 2019 film. Encased in Antarctic ice, Emma triggers explosives to free 521-foot-tall Ghidorah: the ORCA's reviving signal does not subdue this creature to her command. The Monarch team deduces that Ghidorah may have been a seemingly evil and most ancient "rival Alpha to Godzilla ... battling for dominance over the other Titans."[25] This prehistoric nightmare, a veritable storm center in itself, issues powerful lightning bolts from the mouths of each of the three draconian heads. When Godzilla severs one of Ghidorah's snakelike necks, radiation emanating from a volcano helps to rapidly regrow this appendage. In key passages of his 2019 novelization, Keyes suggests that because Legendary's Ghidorah does not appear to be of Earth's terrestrial order, it must be a mythical alien creature that "fell from the stars" in primordial times.[26] Therefore, if true, Ghidorah might not strictly qualify as our fated Medean antagonist (per paleontologist Ward), yet regardless remains a summoned reckoning force against which Earth's natural (more Gaian-esque) Titans must conquer in order to assure mankind's immediate survival. In form and body outlines, Ghidorah is perhaps the most dragodinosauroid (per paleontologist Sanz) of Legendary's daikaiju Titans. Ghidorah's insertion into our planet presumably from without (as opposed to within, i.e., from Hollow Earth) is perhaps analogous as to how *Pacific Rim*'s attacking dino-daikaiju enter from within Earth's bowels via a breach in the Anteverse. As mentioned in Chapter Ten, Ghidorah takes on a variant form in a 2018 Toho anime film.

Mothra (Titanus Mosura): Under Legendary's canon, Hope—remaining in mythological Pandora's Box after awful entities escaped to eternally plague mankind—is personified in the beautiful, feminine gossamer-winged, angelic Mothra. She is an insectoid manifestation of Gaia incarnate, likened to a miracle, in abject darkness an answer to our prayers.[27] Mothra's form differs from her early appearances in Toho's 1960s performances, newly including a stinger she uses to vanquish Rodan in the hellish Bostonian climactic battle of *Godzilla: King of the Monsters*. Heroic Mothra—never yielding to Ghidorah's control—sacrifices herself (not unlike both elder and younger Serizawas did so as well), literally rescuing mankind from extinction. Mothra is said to share a sort of symbiotic relationship with now generally good-guy Godzilla. The Queen of the Monsters as dubbed by Keyes.[28]

MUTOs (Massive Unidentified Terrestrial Organisms): These giant bug-like, radioactive creatures were introduced by Legendary in their 2014 feature film (as this term was more specifically applied) *Godzilla*, and were further detailed in Cox's 2014 novelization from the movie script. For more on these most worthy nemeses to Godzilla, as well as their evolutionary relation to the Jinshin Mushi, see Chapters One and Ten in this book.

Nemesis: Jeremy Robinson's series of 2010s novels (Breakneck Media) dealt with extreme, world-shifting horrors of misguided bio-engineering, introducing a DNA-contrived dino-daikaiju in his 2012 novel *Project Nemesis*. The creature (in part derived from DNA of an unfortunate girl named "Maigo") metamorphoses into a gigantic 380-foot-long creature that destroys part of Boston. In sequels—with Washington, D.C., and other cities around the globe in their sights and the world in flames—Nemesis is joined by other enormous and monstrous kaiju products (e.g., Scylla, Scrion, Drakon, Typhon and Karkinos) of biotech that plague mankind and prove worthy adversaries to Nemesis. See Chapter Seven for more.

Otachi: Not a Titan per the Legendary Godzilla mythos, but yet a genetically (evolved) derived 210-foot-long dino-monster daikaiju appearing in 2013's *Pacific Rim*. Otachi tag-teams with Leatherback in fighting man's Jaegers. Given its spiked-tail weapon, Otachi may be more genetically allied to the *Stegosaurus* than other kaiju kin from the Anteverse. This Category-4 kaiju creature unfolds its wings during combat, taking to the air, and has a "second brain" in its tail region.

Paleosaurus: This "paradinosauroid," per Sanz's 2002 definition, is a highly radioactive, amphibian pseudo-sauropod dino-monster that terrorizes London (and abroad) in 1959's *The Giant Behemoth* (Allied Artists, directed by Eugene Lourie). According to Sanz, generally "Paradinosauroids are quadruped animals similar to sauropods but with much shorter necks and with powerful heads akin to those of the great carnivorous dinosaurs."[29] If it had appeared in Legendary's series, Paleosaurus might also qualify as a Titan, but at least in the annals of *kaiju eiga*, it is clearly dino-daikaiju.

Pet: A monster featured in perhaps the first "*kaiju eiga*–ish" film—a short animated production, *The Pet*, created and directed by bronto-monster maven Winsor McCay in 1921. This monster at large, however, isn't a dino-monster, but rather a puppy-like animal that nightmarishly transforms into a huge city-smashing creature worthy of being attacked by a military squadron of biplanes. Rather reminiscent of Ray Harryhausen's stop-motion animated Ymir, appearing in 1957's *Twenty Million Miles to Earth* (Morningside Productions/Columbia, directed by Nathan Juran).[30]

Quetzalcoatlus: Not just a reference to the magnificently large Late Cretaceous airplane-sized fossil pterosaur known from Texas, but also an indeterminate (presumably flying draconian) kaiju mentioned briefly in Keyes's 2021 novelization.

Reptilicus: This often derided 100-foot-long, venomous, slime-spewing, winged creature featured in its 1961 titular film (Alta Vista Productions, script by director Sidney Pink and Ib Melchoir) constitutes a quasi-dino-monster, possibly of the dragodinosauroid persuasion.[31]

Rhedosaurus: Like the lesser known Paleosaurus, another paradinosauroid—stop-motion animated by Ray Harryhausen—first appearing in 1953's *The Beast from 20,000 Fathoms*. (See Chapter Eleven for more perspective on this film.) In Godzilla lore, Rhedosaurus's cinematic successes in 1953 helped pave the way toward Godzilla's 1954 debut. (See Figure 24.)

Rodan: In Legendary's 2019 film, prehistoric winged Rodan, the "Fire Demon," erupts from a volcano aptly named "Nest of the Demon" on La Isle de Mara. Slathered in lava as it emerges, soon sonic boom–inducing Rodan answers Ghidorah's calling. An old Toho classic movie favorite mainstay (and the 3rd of Toho's prehistoric dino-monster/daikaiju clan), ultimately Mothra nearly kills Rodan. Keyes further infuses formidable foe Rodan with a persona differing rather from Toho's former staged appearances of this monster, stating that "Satan himself could not be as terrible." Also, "Rodan *was* a fire demon, carrying the blaze with him wherever he went, just as Godzilla had his blue radiation and Monster Zero his golden-lightning."[32]

Shinomura: A gigantic bat-like creature, menacing Godzilla since the time of America's Castle Bravo hydrogen bomb testing in 1954. See Chapter Ten.

Skull Crawlers: Kong's ancient nemeses from within Hollow Earth that occasionally transgress onto the surface of Skull Island in 2017's *Kong: Skull Island*. These gigantic reptilians give Kong reason to protect the planet from his home base, Skull Island. See Chapter Six for more on these denizens of Legendary's Monsterverse. (See Figure 19.)

Slattern: The first known Category-5 kaiju from *Pacific Rim*'s Anteverse. Six-hundred-foot-long Slattern, brimming with toxicity and imbued with intelligence, breached in 2025. (Just like the propensity for climate change to grow stronger and more extreme if causal circumstances are left unabated, kaiju invading through the breach continue to become larger and more powerful.) Its face is shaped like that of a hammerhead shark. Equipped with three-clawed hands on its long arms and three armor-piercing tails, and invulnerable to extremely high-water pressure, it proves a formidable foe to Earth's inhabitants. Its name appears in Irvine's 2013 novelization, but not in the 2013 film. This most imposing creature nearly thwarts man's effort to close the breach.

Spinosaurus: A real genus of carnivorous 40-foot-long theropod, African Late Cretaceous dinosaur, little known until more recently, especially following the discovery of the related *Suchomimus* during the late 1990s. This dinosaur distinctively sported a large "sail" along its vertebral column. The likely morphology of *Spinosaurus* in scientific restorations has shifted in

Figure 30: A 1:35-scale model sculpture, crafted by PNSO of China by Zhao Chang and Yang Yang, showing the Spinosaurus in modern (2016) guise (author's collection).

recent years—as for example reflecting its aquatic nature and dietary habits being possibly restricted to piscivory. But movie monster fans will recall that *Spinosaurus* defeated a *Tyrannosaurus* in *Jurassic Park III* (Universal/Amblin, 2001, directed by Joe Johnston).

Figure 31: Frank Bond's 1899 restoration of the Stegosaurus, with plates flattened in a turtle-like carapace over its spine and dorsal areas, and with too many spikes covering its back and tail. Doesn't this hypothetical portrayal rather resemble Toho's Anguirus? (From Charles Whitney Gilmore, "Osteology of the Armored Dinosauria in the United States National Museum, with a special reference to the genus *Stegosaurus.*" *Memoirs of the United States National Museum, 89,* pp. 1–316.) Compare this image with Figure 32.

Stegosaurus: One of paleontology's most recognizable dinosaurs, quadrupedal *Stegosaurus* is that familiar genus with rows of bony plates protruding from its spine. Usually regarded as a creature of

Figure 32: A model kit representation showing a modern perspective on Stegosaurus—one of three key dinosaurs resulting in Godzilla's chimeric fusion. This sculpture is based on a design for Aurora toys, although later manufactured by Lunar Models during the 1990s (author's collection).

gentle disposition, over a century, this dinosaur has appeared numerous times in dino-monster movies. For instance, in 1933's *King Kong* a fierce, charging (giant!) *Stegosaurus* is the first dino-monster encountered by Ann Darrow's rescue party on Skull Island—beyond the ancient Great Wall. And, of course, it is *Stegosaurus's* osteoderms (e.g., plates) that became, in an artistic sense, incorporated into distinctive Godzillean irradiative fins during Toho's suitmation design. In Legendary's lore, *Stegosaurus* is genetically ancestral to *Pacific Rim's* kaiju Otashi.

Titanus Amhuluk: A gigantic, seaweedy-looking, viny creature that battles yet loses to Godzilla in a 2021 graphic novel after the great showdown with Ghidorah in Boston.

Titanus Behemoth (Mapinguary): A double-tusked, bipedal woolly mammoth variety of Titan (but anatomically) "built more like a giant ground sloth,"[33] which made a cameo in Legendary's 2019 movie *Godzilla: King of the Monsters.* This long-armed, clawed god-Titan was preserved in a Brazil cave since the dawn of civilization. Per a 2021 graphic novel, in the wake of Ghidorah's destruction in Boston, Godzilla subdued Behemoth.

Titanus Methuselah: In Keyes's 2019 novelization, Kraken unearths itself from a tree-covered, mountain-sized pile of rocky soil in Germany. This quadruped creature "had a face and horns like a bull from some ancient hell."[34] It makes a cameo in Legendary's 2019 film, and has been referred to as one of the oldest Titans.

Titanus Scylla: This giant, spidery, chimeric kaiju creature, further equipped with tentacles, appears in a cameo in Legendary's 2019 film, emerging in Sedona, Arizona. In the 2021 film, Kong wins a skirmish versus this creature within Hollow Earth.

Titanus Tiamat: This serpent-like kaiju is referred to and illustrated in Legendary's 2021 graphic novel *Godzilla: Dominion*, the prequel to the 2021 film *Godzilla vs. Kong.*

Trespasser: The first of the kaiju to invade Earth, ascending from the Anteverse via a breach in a hydrothermal vent on the Pacific Ocean floor in August 2013—triggering the wave of *Pacific Rim's* derived dino-daikaiju. Trespasser is adorned with a knifelike, crested skull. It took three nuclear missiles to destroy this horrifying harbinger of what was to come.

Triceratops: "Three-horned-face" is one of our most famous, distinctive and quintessential dinosaurs, in conjunction with its ancient foe— *Tyrannosaurus*—personifying, more so than any other dino-duo, the savagery of primeval times. This pairing of Late Cretaceous, North American dinosaurs was imaginatively captured in two of Charles R. Knight's most magnificent painted restorations of the early 20th century, thus indelibly leaving its mark on innumerable dinosaur aficionados. In over a century, *Triceratops* also has made many appearances in film as a science fictional mainstay, as well as in sci-fi and horror literature—too many such occurrences to mention here. Generally not a dino-daikaiju, yet one of man's most familiar and treasured dino-monsters. Most readers might not realize that in the 1932 *King Kong* novelization by Delos W. Lovelace, Kong battled and battered a herd of three-horns! Individuals of this genus of dinosaur may have witnessed (seen and heard) the deafening, cataclysmic era-ending fall of the asteroid 66 million years ago—paving the way for mankind's eventual, yet not foreordained ascendance.

Tyrannosaurus: Especially, perhaps, since its titanic battle with Kong on Skull Island in the 1933 RKO classic film, *Tyrannosaurus* has become the most popular dinosaur ... and *real* dino-monster of them all! Not a true kaiju by any means, though. Often paired with *Triceratops* in Mesozoic lore, thanks to Knight's mesmerizing century-old portrayals.

Uuangi-Ni: In the wake of 1963's Americanized *King Kong vs. Godzilla*, Charlton's comic series, *Konga*, introduced a giant, bipedal, horned and spiky volcanic dino-monster (resembling Gigantis from an American advertisement for the 1959 titular film) that battles the giant ape to the death in "Konga Battles the Creatures of Uuang-Ni!"[35] Konga must defeat a pair of these fiery pseudo-theropods intruding from the bowels of Earth in order to save the island and its native people. This is not unlike the circumstances of how Kong strove to keep the subterranean Skull Crawlers in check in Legendary's 2017 film.

Varan: A giant, spiky dino-monster most known to western movie fans in Toho's Americanized 1958 film *Varan the Unbelievable*.

Velociraptor: Not a daikaiju due to its meager (slightly less than human) size, but, like many of the monsters appearing in 1933's *King Kong*, it was enlarged to the size of a related real dinosaur genus—*Utahraptor*—for the purposes of thrilling audiences in 1993's *Jurassic Park* and other sequels and in 2015's *Jurassic World* (and sequel). Arguably, this is one of the fiercest, blood-thirstiest and most cunning of any real dino-monster in the movies. This named genus became so popularized principally because of *Jurassic Park*'s popularity. *Velociraptor* DNA and genes, obtained from fossil insects preserved in amber, become essential ingredients in the laboratory formulation of both Indominus and Indoraptor.

Zigra: A semi-aquatic, alien kaiju from Planet Zigra, featured in 1971's *Gamera vs. Zigra*. See the Epilogue for more on this creature.

Chapter Notes

Preface

1. Gillian Anderson cited in a 2016 interview, published online (GQ.com/story/gillian-anderson-x-files). "Gillian Anderson Answers All Your Questions About the New X-Files Revived." Jan. 19, 2016, by Lauren Larson. Also cited in Allen A. Debus, *Dinosaur Memories II: Dino-Daikaiju & Paleoimagery*, Chapter Twenty-four, "Reflections of Doomsday: When did Dinosaurs become Apocalyptical" (Createspace, 2017), p. 261.

2. Rene Dubos quoted in Garrett's *The Coming Plague: Newly Emerging Diseases in a World Out of Balance* (New York: Penguin Books, 1994), p. 13.

3. In fact, thematic origins of modern paleoart, providing "an important precedent for some of the earliest scenes from ... deeper, *pre-human* time. ... The tradition of biblical illustration would serve as an important model for scenes from prehumen history," may be witnessed in early 18th-century paintings expressing biblical scenes, such as John Martin's (1789–1854) *The Deluge* (1828) or *The Fall of Babylon* (1819). Martin is also recognized as one of the earliest paleoartists for his painted visions of several prehistoric scenes. See Martin J. S. Rudwick's *Scenes From Deep Time: Early Pictorial Representations of the Prehistoric World* (Chicago: University of Chicago Press, 1992), pp. 20–24, 79–85.

4. Ban Ki-Moon, "How Covid-19 could aid global relations," cited in *Time*, vol. 195, no. 15 (April 27–May 4, 2020), p. 56.

5. Dutch historian Rutger Bregman now speculates whether "COVID-19 seems like ... the prequel to the global climate crisis. ... Now is the moment to change the world," "The moment to change the world is right now," in *Time*, vol. 195, no. 18 (May 18, 2020), p. 35.

6. Particularly since the early 1980s, popular examination of paleoart and study of historical paleoartists has certainly come into vogue. Recently I identified perhaps the earliest (1870) printed use of the term "paleo-artist," as explained further in my 2019 book *Prehistoric Monster Mash: Science Fictional Dinosaurs, Fossil Phenoms, Paleo-pioneers, Godzilla & Other Kaiju-saurs* (Amazon Kindle), pp. 224, 240.

7. Man has had a long-standing preference for the subliminally psychologically pleasing pairing of phantasmagoric prehistoric pugilists—that is, imagined terrible conflict between the greatest and most formidable creatures representing their former ages and times. During the late 1790s, Thomas Jefferson (a paleontologist in his own right) envisioned the tumultuous battle that would have been witnessed in archaic times between "Great Claw" and the American "Incognitum," (or in today's parlance, *Megalonyx* versus Mammoth). A few decades later, early paleoartists and writers (e.g., Jules Verne) portrayed marine warfare between *Ichthyosaurus* and *Plesiosaurus*—a popular mid-19th-century theme. Meanwhile, paleontologists avidly imagined terrestrial combatants, *Iguanodon* versus *Megalosaurus*. And 1870s

editions of Jules Verne's novel *Journey to the Center of the Earth* also contain passages (written by an unknown author) pitting the prehistoric, unevolved Ape Gigans versus the Shark Crocodile. By the early 20th century, Charles R. Knight had immortalized the ritualistic savage struggle between *Triceratops* and *Tyrannosaurus* in two museum paintings—still perhaps the most distinctive dinosaurian paleoart theme. And then via RKO's classic 1933 film *King Kong*, we now also may add the prehistoric ape versus dino-monster to the popular pantheon. (See Debus, *Dinosaur Memories II, op. cit.*, pp. 275, 283–288; Debus, *Prehistoric Monsters: The Real and Imagined Creatures of the Past That We Love to Fear* (Jefferson, NC: McFarland and Company, Inc., Publishers, 2010).

8. Allen A. Debus Foreword (pp. 1–2) in Mike Bogue's *Apocalypse Then: American and Japanese Cinema, 1951–1967* (Jefferson, NC: McFarland and Company, Inc., Publishers, 2017).

9. For more on hope and optimism at the 1939 World's Fair, see David Gelernter's book *1939: The Lost World of the Fair* (New York: The Free Press, 1995).

10. Another contemporary pop-cultural reference is heard in lyrics to The Guess Who's 1970 hit song, titled "No Sugar Tonight: It's the New Mother Nature Taking Over."

11. One striking example of the Late Cretaceous-Paleogene (or "K-T") mass extinction event of 66 million years ago, presented in a science-fictionalized (time travel) format, may be enjoyed in Will Hubbell's novel *Cretaceous Sea* (New York: Ace Books, 2002). Along with many metaphorical sci-fi examples, I have extensively discussed the real variety of dinosaurian catastrophes in my 2016 book, *Dinosaurs Ever Evolving, op. cit.* Chapter Thirteen in that volume, titled "Prehistoric Life Spawns an Environmental Movement," might be of some further interest to dino-phile readers.

12. Sean Rhoads and Brooke McCorkle, *Japan's Green Monsters: Environmental Commentary in Kaiju Cinema* (Jefferson, NC: McFarland & Company, Inc., Publishers, 2018).

Introduction

1. A. Debus, *Dinosaur Memories II: Pop-cultural Reflections on Dino-daikaiju & Paleoimagery* (Amazon-Createspace, 2017), p. 276.

2. Ted Okuda and Mark Yurkiw, *Chicago TV Horror Movie Shows: From Shock Theater to Svengoolie* (Chicago: Lake Claremont Press, 2016).

3. Darlene Geis, *Dinosaurs and Other Prehistoric Animals* (New York: Grosset & Dunlap, Inc., 1959).

4. Ted Johnson, "Godzilla: King of the Good Guys," in *G-Fan*, vol. 1, no. 122 (Winter 2018), pp. 56–60.

5. This theory wasn't proposed on basis of intriguing geochemical and geological evidence until 15 years later in 1980.

6. Loren Eiseley, *The Invisible Pyramid* (New York: Charles Scribner's Sons, 1970).

Chapter One

1. Carl Sagan. *The Dragons of Eden: Speculations on the Evolution of Human Intelligence* (New York: Ballantine Books, 1977), p. 160.

2. The term "dragodinosauroid" was coined by paleontologist Jose Luis Sanz in his book *Starring T. Rex!: Dinosaur Mythology and Popular Culture* (Bloomington: Indiana University Press, 2002), pp. 109, 115.

3. Earth systems science will be further defined, however. Think of generally as the old Earth Science curriculum, although with added, harder-leaning, coupled emphasis on biogeochemistry, geophysics and environmental chemistry. The concept of a Gaian Earth hinges on the chemical disequilibrium noted in our ocean-atmosphere system, one of low entropy, characterized by presence of living things. For example, components of our atmospheric gases are unstable, chemically reactive, and therefore incompatible. If oxygen's concentration at sea level (currently 21%) increased, say, by another 10% to 15%, wildfires might be raging continually until all carbon-bearing plants were incinerated to carbon dioxide. Biodiversity is key to

maintaining a proper oxygen level fit for the Anthropocene Era. Garrett, *Coming Plague*, p. 555. Earth's conditions, regulated via interplay between biota and the environment (both of which evolve in tandem through time) have remained suitable for the continuation of some forms of life for over 4 billion years, despite an onslaught of ravaging geological ages of pollution and environmental upset-causing extinctions (some of extraterrestrial origin) noted at several horizons in the fossil record. Negative feedbacks have prevented ecological conditions from reaching (high-entropy) chemical equilibrium, such as on Venus, where due to loss of hydrogen gas to space (see Lovelock's *Ages of Gaia*, W. W. Norton & Company, 1988, p. 86) a runaway "greenhouse effect" resulted in a lifeless planet. An uncontrolled, unmediated series of positive feedbacks can create havoc for life. For considerably more in context of modern Earth Systems Science, see *The Earth System, 3rd ed.* (Pearson Education-Dorling Kindersley, 2016), by Lee R. Kump and James F. Kasting.

4. John Gribbin and Mary Gribbin, *He Knew He Was Right: The Irrepressible Life of James Lovelock* (London: Penguin Books, 2009), p. xiv.

5. Peter S. Alagona. Book review of item in note 4 above, in *Isis* (vol. 101, no. 1, March 2010), p. 254; Of the New Age connections, Lovelock has stated, "The first paper to mention (Gaia) was published in the *Proceedings of the American Astronautical Society* in 1968. The title … was 'Planetary Atmospheres: Compositional and other changes associated with the presence of Life.' It was an almost unnoticed paper. … So the Gaia concept was born at the peak of the New Age—contemporary with Woodstock and the Beatles, which perhaps accounts for why so many scientists still regard it as part of the plethora of New Age nonsense that was around at the time. But not all of us were hippies with our rock chicks." Lovelock, *The Vanishing Face of Gaia: A Final Warning* (New York: Basic Books, 2009), pp. 159–160.

6. Peter Ward, *The Medea Hypothesis: Is Life on Earth Ultimately Self-Destructive?* (Princeton, NJ: Princeton University Press, 2009), p. xx. According

to Edith Hamilton's *Mythology*, the name Gaia (or "Gaea") referred to Mother Earth, one of the first two living creatures having an "appearance of life." Author William Golding recommended to Lovelock that this name would be aptly applied to the latter's developing theory of our "living" planetary organism. And in Greek mythology, wicked, vengeful Medea facilitated Jason's (of "Argonauts" fame) exploits, yet in the end permitted the death of her own two sons.

7. *Ibid.*, p. 22.

8. *Ibid.*, p. 81.

9. Rhoads and McCorkle, *Japan's Green Monsters*.

10. Max Page, *The City's End: Two Centuries of Fantasies, Fears, and Premonitions of New York's Destruction* (New Haven, CT: Yale University Press, 2008).

11. *Ibid.*, p. 2.

12. Al Gore (writer), *An Inconvenient Truth* (film documentary, 2006), directed by Davis Guggenheim (Lawrence Productions; Participant Productions).

13. Alex Irvine, *Pacific Rim (The Official Movie Novelization)* (London: Titan Books, 2013), p. 24; Allen A. Debus, "Toward a Unified Theory of Dinosaurian Kaiju: A *Pacific Rim*-inspired, 'what if'?" Chapter Twenty-five in *Dinosaur Memories II: Dino-daikaiju & Paleoimagery* (Createspace, 2017), pp. 266–269.

14. Greg Keyes, *Pacific Rim: Uprising-Ascension* (London: Titan Books, 2018), p. 75.

15. *Ibid.*, p. 313.

16. Mary Shelley, *Frankenstein, or the Modern Prometheus* (New York: Pyramid Books, 1818; 1957 reprint); Michael Crichton, *Jurassic Park* (New York: Ballantine Books, 1990), p. 313.

17. Sam Enthoven, *TIM: Defender of the Earth* (London: Random House, 2008), p. 3.

18. Elizabeth Kolbert, *The Sixth Extinction: An Unnatural History* (Henry Holt and Company, 2014).

19. See, for example, Michael Allaby's and James Lovelock's *The Great Extinction: The Solution to One of the Great Mysteries of Science—The Disappearance of the Dinosaurs* (Garden City, NY: Doubleday & Company, Inc., 1983), pp. 170–174.

20. James Lovelock, *Gaia: A New Look*

at Life on Earth (Oxford University Press, 1979; 1987 reprint), p. 43. If man's bioengineering impacts an "ordinary" course of evolution, then wouldn't the outcome be "natural" anyway ... or alternatively somehow wrong? A philosophical matter for readers to ponder, but beyond scope of this book.

21. While this is a quote from *Jurassic World*, the anticipated, analogous thought of eventual boredom with dinosaurs was perceived sixteen years prior by W. J. T. Mitchell in his book *The Last Dinosaur Book: The Life and Times of a Cultural Icon* (Chicago: University of Chicago Press, 1998). While this is a theme of his book, see especially pp. 2, 24–25 in Mitchell's book. Also see my refutation of his claim in *Dinosaur Sculpting: A Complete Guide* (Jefferson, NC: McFarland & Company, Inc., Publishers, 2018), p. 172.

22. In an (indirect) interview conducted with Glen McIntosh, the artist who designed Indominus rex, graciously facilitated and transmitted by Mike Fredericks on Feb. 18, 2021 (pers. comm.), McIntosh indicated how this newest of unnatural filmic dino-monsters owes greatly to recent discoveries in the paleontological sciences. Concerning how Indominus was conceived he wrote, "I knew that it was going to fight the *T. rex* at the end of the movie so it had to be visually distinct and yet just as iconic." McIntosh continued, "As far as the head shape it was based on *Allosaurus, Majungasaurus, Giganotosaurus, Carnotaurus,* and *Rugops*." Note that four of these theropod genera have been discovered during recent time—the 1980s and 1990s. Furthermore, he wrote,

> You can ... see how similar the *Rugops* skull is to Indominus. This is also why Indominus has a convex shape to the nose and muzzle. Just like the abelisaurs, thick osteoderms and spiky scutes were going to give the skin an interesting texture but all that detail was also going to show off the scale and size of the "Indominus-Rex." Just like an allosaurid, I had the horns on the top of the head pre-orbital. From the front this would help make it look more demonic ... but moved the horns over the eyes like a

Carnotaurus which is why the Legacy effects maquette sculpture is different from the final design and more like an *Allosaurus*. They matched my earlier drawing scales and all! A combination of features also lent to the idea that the Ingen geneticists combined DNA from various theropods to get a sort of a "Frankenstein's dinosaur." You ultimately ... sympathize with the dinosaur because it was raised alone but that's why it's kind of sad and why it really only goes on a rampage for one full day before being brought down. This is also why it had an underbite and the teeth didn't interlock perfectly because it was sort of a genetic mishmash. It looks scary but it probably had to cope with a lot of pain which helped when animating it and giving it character. As far as various strands of DNA being combined, that was completely consistent with the questions I had for paleontologist Jack Horner at the beginning of production. He said that would absolutely be possible so getting his green light was great!

As for the name, McIntosh mentioned that an early suggestion was made for christening the chimera "Diabolus Rex," Later they thought up "Malusaurus," but eventually arrived at Indominus rex— or the "Indomitable king." "The king of something impossible to defeat."

23. This famous and arguably Gaian quote is of course borrowed from 1993's *Jurassic Park*, as uttered by chaotician Ian Malcolm, played by Jeff Goldblum.

24. D. G. Valdron, "Fossils," in *Daikaiju!: Giant Monster Tales* (Australia, University of Wollongong NSW: Agog Press, 2008), pp. 181–195.

25. The first quote in this paragraph is borrowed from Lovelock, *Gaia: A New Look at Life on Earth*, p. 16. The second quote in this paragraph is borrowed from Lovelock, *The Ages of Gaia*, p. xvii.

26. Valdron, *op. cit.*, p. 191.

27. Greg Cox. *Godzilla (The Official Movie Novelization)* (Titan Books, 2014), p. 129.

28. *Ibid.*, p. 256.

29. *Ibid.*, p. 284, emphasis added.

30. James Lovelock, *Ages of Gaia*, p. 153.

31. Ward, *op. cit.*, p. 35.

32. Cox, *op. cit.*, p. 243.

33. Lovelock, *Ages of Gaia, op. cit.*, p. 159. It may also be the case that several similar yet contrasting Gaian philosophies—not simply one consistently—are projected throughout Legendary's 21st-century Monsterverse and allied dino-monster movies.

34. Cox, *op. cit.*, p. 230.

Chapter Two

1. See Chapter Thirty-seven, "A Paleoart Revolution," in my book *Prehistoric Monster Mash*, pp. 423–428; and Valerie Bramwell and Robert M. Peck, *All in the Bones: A Biography of Benjamin Waterhouse Hawkins* (Philadelphia: The Academy of Natural Sciences, Special Publication 23, 2008).

2. Jules Verne, *Journey to the Center of the Earth*. 1867. Reprint trans. William Butcher, (Oxford: Oxford University Press, 1998); Allen A. Debus, "Reframing Verne's Paleontological *Journey*," *Science Fiction Studies*, vol. 33, no. 100, Part 3 (November 2007), pp. 405–420.

3. Allen A. Debus, *Dinosaur Memories II: Dino-Daikaiju & Paleoimagery* (Createspace, 2017), Chapter Twenty-seven, "Two Mysterious Monsters of Early Science Fiction Literature."

4. Martin J. S. Rudwick includes several examples in his book *Scenes From Deep Time: Early Pictorial Representations of the Prehistoric World* (Chicago: University of Chicago Press, 1992).

5. Camille Flammarion, *Le Monde Avant La Creation De L'Homme: Origines de la terre. Origines de la vie. Origines de l'humanite.'* (Paris: C. Marpon and E. Flammarion, 1886). For more on this chimeric stegosaur, see A. Debus, *Dinosaur Memories II, op. cit.*, pp. 228–229 and 231–233.

6. "Most Colossal Animal Ever on Earth Just Found Out West," *New York Journal*, Dec. 11, 1898, p. 1. (byline absent)

7. Ulrich Merkl, *Dinomania: The Lost Art of Winsor McCay, the Secret Origins of King Kong, and the Urge to Destroy New York* (2015).

8. Quotes and suggestions from Merkl, *op. cit.*, pp. 105, 262–263.

9. Donald F. Glut, *Dinosaur Scrapbook* (Secaucus, NJ: Citadel Press, 1980), pp. 200, 229.

10. Allen A. Debus, *Prehistoric Monsters: The Real and Imagined Creatures of the Past That We Love to Fear* (Jefferson, NC: McFarland and Company, Inc., Publishers, 2010), pp. 242–246.

11. Stephen Czerkas, *Major Herbert M. Dawley: An Artist's Life—Dinosaurs, Movies, Show-Biz & Pierce-Arrow Automobiles* (The Dinosaur Museum, 2016).

12. Donald F. Glut, *Jurassic Classics: A Collection of Saurian Essays and Mesozoic Musings* (Jefferson, NC: McFarland and Company, Inc., Publishers, 2001), pp. 174–183.

13. Michael Klossner, *Prehistoric Humans in Film and Television* (Jefferson, NC: McFarland and Company, Inc., Publishers, 2006), p. 243.

14. *Ibid.*

15. Donald F. Glut, *Classic Movie Monsters* (Metuchen, NJ: The Scarecrow Press, 1978).

16. Steve Ryfle, *Japan's Favorite Mon-Star* (Toronto: ECW Press, 1998), p. 81.

17. John LeMay, *Kong Unmade: The Lost Films of Skull Island* (Amazon: Bicep Books, 2019), p. 134.

18. Ryfle, *op. cit.*, p. 27.

19. *Ibid.*

20. *Ibid.*

21. Ed Godziszewski, "The Making of Godzilla," in *Japanese Giants: The Magazine of Japanese Science Fiction and Fantasy*, vol. 1, no. 10, Sept. 2004, pp. 5–47.

22. Ryfle, *op. cit.*, p. 63.

23. *Ibid.*, p. 84.

Chapter Three

1. Isaac Asimov and Frederik Pohl, *Our Angry Earth* (New York: Tor Books, 1991), p. 14.

2. Hutton quoted in Dorion Sagan's *Biospheres: Metamorphosis of Planet Earth* (New York: Bantam Books, 1990), p. 58. For considerably more on Hutton, see Stephen Baxter's *Ages in Chaos: James Hutton and the Discovery of Deep Time* (New York: Forge Books, 2003); Lovelock often mentions the substantial contributions of Lynn Margulis, particularly her

comprehension of the role of microorganisms throughout planetary history in
transforming the composition of Earth's
ocean-atmosphere system.

3. Gribbin, *He Knew He Was Right*,
p. 49.

4. Dorion Sagan, *op. cit.*, p. 46.

5. There were others whose scien-
tific/naturalistic thinking bordered on
Gaian philosophy as well, such as Lars
Sillen (1916–1970), Lawrence J. Hender-
son (1878–1942), Alfred James Lotka
(1880–1912), Henry David Thoreau
(1817–1862), Victor Goldschmidt (1888–
1947) and Rachel Carson (1907–1964).

6. James Lovelock, *The Ages of Gaia:
A Biography of Our Living Earth* (New
York: W. W. Norton & Company, 1988),
p. 19. Here Lovelock further states that
"Gaia is not a synonym for the biosphere. ... Gaia, as a planetary being, has
properties that are not necessarily dis-
cernible by just knowing individual species or populations of organisms living
together."

7. *Ibid.*, pp. 23, 63.

8. James Lovelock, *Gaia: A New Look
at Life on Earth* (Oxford University Press,
1979, 1982 ed.), p. 82.

9. *Ibid.*, p. 92.

10. Asimov and Pohl, *op. cit.*, pp.
17–18.

11. Asimov and Pohl, *Ibid.*, p. 24;
Lovelock further stated in 2009, "if we
fail to curb global heating, the planet
could massively and cruelly cull us, in
the same mercilessly way that we have
eliminated so many species by changing the environment into one where
survival is difficult ... before we start
geo-engineering, we have to ask: Are
we sufficiently talented to take on what
might become the onerous task of keeping the Earth in homeostasis? Consider what might happen if we start by
using a stratospheric aerosol to amelio-
rate global heating—even if it succeeded
it would not be long before we faced the
additional problem of ocean acidification. This would need another medicine,
and so on." Lovelock, *Vanishing Face of
Gaia*, p. 156. Interestingly, paleontologist Peter Ward notes, "As far back as
1996, the official website of the Burning
Man Festival featured Gaia as 'The Cruel
Mistress.'" *The Medea Hypothesis*, p. 25.

Justin Worland worries in *Time* that "climate change is taken for granted—and
we are all left to play a game of survival of
the fittest. Such a game wouldn't just pit
countries against one another—it would
also pit humans against the planet. And
the planet would win every time." "Forces
of Nature," *Time*, April 26/May 3, 2021,
vol. 197, nos. 15–16, p. 66.

12. The term "population bomb" refers
to an out-of-control population increase
of the human race, a consequence which
became feared during the early 1970s
due to Earth's limited carrying capacity.
Paul R. Ehrlich (with Anne Ehrlich), *The
Population Bomb* (Sierra Club/Ballan-
tine, 1968); Arthur N. Strahler and Alan
H. Strahler, *Environmental Geoscience:
Interaction Between Natural Systems
and Man* (Santa Barbara, CA: Hamilton
Publishing Company, 1973), p. 498. As
biochemist Isaac Asimov and Frederick
Pohl note in regard to a "zero population
growth" ideal, "The cause of the immediate crisis isn't sheer numbers, it is the
unrestrained and wasteful use of energy
and resources ... what those people *do*."
Our Angry Earth (New York: Tor, 1991),
p. 24.

13. Dorion Sagan, *op. cit.*, pp. 82–83.

14. Lovelock, *Gaia: A New Look at Life
on Earth*, p. 140.

15. Melissa Goding, "The Honeybee
Whisperers," *Time*, vol. 195, May 18,
2020, p. 41.

16. Lovelock, *Ages of Gaia*, p. 41.

17. *Ibid.*, p. 76.

18. Lovelock, *Gaia: A New Look at Life
on Earth*, pp. 86–87.

19. *Ibid.*, p. 192.

20. Robert A. Berner, *The Phanerozoic Carbon Cycle* (Oxford: Oxford
University Press, 2004). Another indication that Gaia exerts a life-sustaining
force even when her systems are under
extreme duress, during throes of a mass
extinctions event triggered by a sudden extraterrestrial cause, may be noted
in the asteroid impact "kill curves" created by David M. Raup and C. Wylie
Poag, respectively in 1991 and 1999.
Raup derived a function portraying species killed versus mean geological time
between such drastic events, related
to another mathematically derived
Extinction-Impact Curve based on data

showing species killed versus impact crater diameter. Raup, *Bad Extinction: Genes or Bad Luck?* (New York: W. W. Norton & Company), pp. 85, 171. Raup's Extinction-Impact Curve—presuming that all mass extinctions are caused by asteroid/comet impacts—would indicate that for a 50-mile crater caused by a (rather sizeable) 3-mile diameter bolide impacting Earth that one would expect a resultant 35% species kill recorded in the geological record. But Poag has found evidence that in the case of a 50-mile diameter impact crater caused by such a bolide striking the Late Eocene North American coastline 35 million years ago, further exacerbated by yet a *second* similarly-sized contemporaneous impact in Siberia, there is no apparent mass extinction dating from that time. Poag, *Chesapeake Invader: Discovering America's Giant Meteorite Crater* (Princeton, NJ: Princeton University Press), p. 101. Accordingly, to me, Poag's adjustment of Raup's kill curve suggests that perhaps life finds a way whenever it can, exerting to the utmost Gaia's remarkable resilience even in aftermath of such hellish conditions, created by occasional random extraterrestrial causes, to mollify the blow. See also note no. 33 in Chapter Six.

21. Peter D. Ward, *Out of Thin Air: Dinosaurs, Birds, and Earth's Ancient Atmosphere* (Washington, DC: Joseph Henry Press, 2006), p. 45.

22. *Ibid.*

23. *Ibid.*, pp. 180–194.

24. Asimov and Pohl, *op. cit.*, pp. 18–19.

25. Ward, *Medea Hypothesis*, pp. 126–127.

26. Ann Druyan captured essence of this primordial biogeochemical interplay eloquently in her *Cosmos: Possible Worlds* (National Geographic, 2020), pp. 97–100.

Methane is a powerful greenhouse gas, and back then it was the main thing keeping the planet warm. But once again, the oxygen produced by life shook things up. It gobbled up the methane, but it excreted carbon dioxide, a much less potent greenhouse gas—meaning it was not as efficient at trapping heat in Earth's

atmosphere. Earth grew colder and the green life on the land began to die. ... Earth was a snowball, completely encased in snow and ice. The cyanobacteria had gone a little too far. The dominant life-form on the planet came close to total self-annihilation. A sobering thought for beings who happen to occupy that ecological niche today.

27. Peter D. Ward, *Under a Green Sky: Global Warming, the Mass Extinctions of the Past and What They Can Tell Us About Our Future* (Smithsonian Books, 2007).

28. Lovelock quoted in Michael Ruse, *The Gaia Hypothesis: Science on a Pagan Planet* (Chicago: University of Chicago Press, 2013), p. 35.

29. Ruse, *op. cit.*, p. 189.

30. Richard Dawkins, *The Extended Phenotype: The Long Reach of the Gene* (Oxford: Oxford University Press, 1982), 236; Lovelock's rebuttal to Dawkins may be found in his *Vanishing Face of Gaia, op. cit.*, pp. 169–170, 193. Here Lovelock stated, "To me it was obvious Richard Dawkins's pure biology and the geochemist's pure chemistry were unable to explain the Earth. ... Gaia ... fails the biologist's test because it does not reproduce, nor can there be natural selection among planets." Dorion Sagan might disagree with that statement, though.

31. A major theme of Dorion Sagan's *Biospheres.*

32. Eiseley, *Invisible Pyramid.*

33. Rachel Carson, *Silent Spring* (Greenwich, CT: Fawcett Crest, 1962), p. 186.

34. Garrett, *Coming Plague.*

35. George Gaylord Simpson in 1957, quoted in Rachel Carson, *op. cit.*, p. 189.

Chapter Four

1. Monte Reel, *Between Man and Beast: An Unlikely Explorer, the Evolution Debates, and the African Adventure That Took the Victorian World by Storm* (Doubleday, 2013), p. 53.

2. *Ibid.*, p. 117.

3. Paul A. Woods, ed., *King Kong Cometh* (Plexus Publishing, Ltd., 2005), p. 8.

4. Mitchell, *Last Dinosaur Book*, pp. 201–205, quote borrowed from p. 252.

5. Paul Du Chaillu. *Explorations and Adventures in Equatorial Africa* (1861, Ostara Publications.com, 2013), p. 59.

6. *Ibid.*

7. *Ibid.*, p. 71.

8. *Ibid.*, p. 299.

9. Quotes in this par. borrowed from Du Chailu, *op. cit.*, pp. 70–71, 299, 360, and Reel, *op. cit.*, pp. 79, 126.

10. Bramwell and Peck, *All in the Bones*, p. 76.

11. Stephen J. Gould, "Knight Takes Bishop?" in *Bully For Brontosaurus* (New York: W. W. Norton, 1991), pp. 385–401; Adrian J. Desmond, *Huxley: From Devil's Disciple to Evolution's High Priest* (Perseus Books, 1994, 1997); Stephen J. Gould, "A Visit to Dayton," in *Hen's Teeth and Horse's Toes: Further Reflections in Natural History* (New York: W. W. Norton, 1991), pp. 263–279.

12. Jules Verne, *A Journey to the Center of the Earth* (1864, c. 1872, Signet Classic, 1986).

13. Arthur Conan Doyle, *The Lost World* (London, 1912; 1925 ed., New York: A. L. Burt Company, Publishers), reprinted in *The Annotated Lost World: The Classic Adventure Novel by Sir Arthur Conan Doyle In a New Carefully Referenced and Lavishly Illustrated Edition*, edited by Roy Pilot and Alvin Rodin (Indianapolis: Wessex Press, 1996).

14. For more on these and other vintage, relevant titles, see Allen A. Debus, *Dinosaurs in Fantastic Fiction: A Thematic Survey* (Jefferson, NC: McFarland & Company, Inc., Publishers, 2006). Also see Ulrich Merkl's, *Dinomania: The Lost Art of Winsor McCay, the Secret Origins of King Kong, and the Urge to Destroy New York* (Fantagraphics, 2015), p. 194, and A. Debus's *Prehistoric Monsters: The Real and Imagined Creatures of the Past That We Love to Fear* (Jefferson, NC: McFarland and Company, Inc., Publishers, 2010), Chapter Five, pp. 87–107.

15. Woods, *op. cit.*, p. 13.

16. That is in Toho's *Godzilla Raids Again* (1955), as Americanized in 1959 and retitled *Gigantis the Fire Monster*.

17. Verne, *op. cit.*, Chapter Forty, c. 1872 ed. and Allen A. Debus, *Dinosaur Memories II: Dino-Daikaiju &* *Paleoimagery* (Createspace, 2017), pp. 278–282.

18. Debus, *Ibid.*

19. Neil Pettigrew, *The Stop-Motion Animation Filmography* (Jefferson, NC: McFarland & Company, Inc., Publishers, 1999), pp. 161–162, 395. Paul Mandell discusses the scripted horned dinosaur scene, which differs in detail from the novelized sequence in *King Kong Cometh*, pp. 84–87.

20. Delos W. Lovelace, *The Illustrated King Kong* (New York: Grosset & Dunlap, Inc., 1976 reprint).

21. Delos W. Lovelace, *King Kong* (1932, New York: The Modern Library, 2005 reprint with Introduction by Greg Bear), p. 89. For more on 'Kong' movie novelizations, see Chapter 18, "Kong versus Godzilla—A Battle of Novelizations," in *Prehistoric Monster Mash*, pp. 187–206.

22. For much more on the old-timey idea of the 'hopping' kangaroo-lizard style of dinosaurian locomotion in sci-fi, see my *Prehistoric Monsters* volume (Jefferson, NC: McFarland & Company, Inc., Publishers, 2010), pp. 184–186.

23. Lovelace, Modern Library ed., *op. cit.*, p. 100.

24. Perhaps the earliest, influential published image of Charles R. Knight's restoration of a snaky-looking elasmosaur was in an 1897 issue of *Century Magazine* Vol. LV.

25. Lovelace, Modern Library ed., *op. cit.*, p. 117.

26. I can't exactly reference the publication or issue of Forrest J Ackerman's publications in which this appeared … but indeed it did and indelibly emblazoned the poem upon my memory.

27. Pettigrew, *op. cit.*, p. 399.

28. Steve Archer, *Willis O'Brien: Special Effects Genius* (Jefferson, NC: McFarland & Company, Inc., Publishers, 1993), p. 18.

29. Don Glut, *I Was a Teenage Movie-Maker: The Book* (Jefferson, NC: McFarland & Company, Inc., Publishers, 2007).

30. Allen A. Debus, *Dinosaur Memories II*, *op. cit.*, pp. 62–75.

31. Woods, *op. cit.*, pp. 158–169.

32. Another Japanese entry was 1933's *Wasei Kingu Kongu* all prints of which

were destroyed during the 1945 Hiroshima bombing. Later, in 1949, a giant gorilla-ape played by suit actor Charles Gemora appeared in Abbott and Costello's *Africa Screams*. John LeMay, *Kong Unmade: The Lost Films of Skull Island* (Bicep Books, 2019, pp. 72–73, 90–97), and Ken Hollings "King Kong Appears in Edo," in *King Kong Cometh, op. cit.,* p. 118.

33. See my *Dinosaur Memories II: Dino-Daikaiju & Paleoimagery,* 2017, pp. 283–288.

34. Factors also included 1953's release of *The Beast From 20,000 Fathoms,* the 1954 Lucky Dragon/Bikini Atoll hydrogen bomb test incident, and of course the two atomic bombings of Japan to end World War II in August 1945. (See Chapter Nine "Nuclear Dragon— Godzilla and the Cold War," in my 2016 book, *Dinosaurs Ever Evolving* (Jefferson, NC: McFarland & Company, Inc., Publishers, 2016) for more.)

35. John LeMay, *Kong Unmade: The Lost Films of Skull Island,* 2019, p. 173.

36. Mark F. Berry, *Dinosaur Filmography* (Jefferson, NC: McFarland & Company, Inc., Publishers, 2002), p. 196. Paul Mandell as well was unimpressed with the 1976 entry, per his 1973 article contribution reprinted in *King Kong Cometh,* see pp. 127–132.

37. Glut quoted in *King Kong Cometh,* p. 142. Donald F. Glut, *Classic Movie Monsters.* Chapter VIII "His Majesty, King Kong," (Metuchen, NJ: Scarecrow Press, 1978), pp. 282–373.

38. Berry, *op. cit.,* p. 278. In *King Kong Cometh, op. cit.,* p. 117, Don Glut mentions another Kong flick of the period, 1967's *Mad Monster Party.*

39. Christopher Golden, *King Kong (The Official Movie Novelization)* (Pocket Star Books, 2005), p. 279.

40. DeVito and Strickland, *Kong: King of Skull Island.* Milwaukie, OR: DH Press, 2004), p. 110.

41. DeVito and Strickland, *King Kong of Skull Island* (DeVito Artworks, 2017), p. 226.

42. Another Kong "versus" dinosauria show should be mentioned. In *Kong: King of the Apes,* a 2018 cartoon spanning 23 episodes, Kong routinely encounters *T. rexes,* which then must be defeated.

Chapter Five

1. Dennis R. Dean, *Gideon Mantell and the Discovery of Dinosaurs* (Cambridge: Cambridge University Press, 1999), p. 131.

2. Donald F. Glut, *Dinosaurs: The Encyclopedia* (Vol. I) (Jefferson, NC: McFarland and Company, Inc., Publishers, 1997), separate entries for *Stegosaurus* (pp. 840–853), *Tyrannosaurus* (pp. 945–960) *and Iguanodon* (pp. 490–500). And while *Stegosaurus* really did not have two brains (its only brain resided within its skull), Toho's kaiju-saur Anguirus was also imbued with a similar pairing of brains in 1955's *Godzilla Raids Again.* Thus, strangely—despite their quite differing bodily morphologies—both Godzilla, and (in part) ankylosaurian/ceratopsian-headed Anguirus may also be conceptually considered as variant stegosaurs.

3. Charles W. Gilmore, *Osteology of the Armored Dinosauria in the United States National Museum, With Special Reference to the Genus Stegosaurus* (U.S. National Museum Bulletin, no. 89, 1914), p. 123.

4. Barnum Brown, "Tyrannosaurus, a Cretaceous Carnivorous Dinosaur," in *Scientific American* (Oct. 9, 1915), pp. 322–323.

5. Mark P. Witton, *Life Through the Ages II: Twenty-first Century Visions of Prehistory* (Bloomington: Indiana University Press, 2020), p. 4.

6. Henry Fairfield Osborn, "A Great Naturalist: Edward Drinker Cope," *The Century Magazine* LV (November 1897): pp. 10–15.

7. Allen A. Debus and Diane E. Debus, Chapter Sixteen: "The Ages of Zallinger," in *Paleoimagery: The Evolution of Dinosaurs in Art* (Jefferson, NC: McFarland and Company, Inc., Publishers, 2002), pp. 108–110, 254–255.

8. Phil Hore, "Deinonychus," *Prehistoric Times* no. 129 (Spring 2019), p. 30.

9. Debus and Debus, *op. cit.,* p. 159.

10. Ed Godziszewski, "The Making of Godzilla," in *Japanese Giants: The Magazine of Japanese Science Fiction and Fantasy,* no. 10, 2004, p. 5.

11. *Ibid.,* p. 6.

12. John LeMay, *Writing Japanese Monsters* (Bicep Books, 2020), p. 11.

13. *Ibid.*

14. The tale of Kayama's involvement in the Godzilla storyline is detailed in Godziszewski, *op. cit.*, pp. 7–11.

15. *Ibid.*, pp. 11–12.

16. David Norman, *Dinosaur!* (New York: Prentice Hall, 1991), pp. 160–169.

17. Both *Iguanodon* and *Stegosaurus* were cast as gigantic (much larger-than-life) curious dino-monsters, peering into early upper-story skyscraper windows during the 1880s in artwork appearing in editions of Camille Flammarion's *Le Monde Avant La Creation De L'Homme* (1886).

Chapter Six

1. Lovelock quoted in Gribbin, *He Knew He Was Right*, p. 128. This concept is the core of Lovelock's Gaian regulatory system.

2. Debus, *Dinosaurs in Fantastic Fiction.*

3. This primitive (yet not edenic), prehistoric balanced ecosystem is suggested in other Kong-related publications of the 2000s as well, such as *The World of Kong: A Natural History of Skull Island* (New York: Pocket Books, Weta Workshop, 2005), Foreword by Peter Jackson. This lavishly illustrated volume was a tie-in to Universal's 2005 film *King Kong*—unrelated to the Legendary movie series.

4. Arvid Nelson and Zid, *Skull Island: The Birth of Kong (The Official Comic Prequel to the Major Motion Picture)* (Legendary Monsterverse, 2017).

5. Allen A. Debus, "Mystery of the Spider-Men," in *Prehistoric Monster Mash*, Chapter Seven, pp. 86–96.

6. Delos W. Lovelace, *King Kong* (1932, New York: The Modern Library, 2005 reprint with Introduction by Greg Bear).

7. Graham Edwards, "A Lonely God," *Cinefex 152*, no. 152 (April 2017), pp. 62, 82.

8. Arvid Nelson, *op. cit.* (There is no page numbering system in this graphic novel.)

9. Tim Lebbon, *Kong: Skull Island (The Official Movie Novelization)* (Legendary, Titan Books, 2017), p. 312.

10. *Ibid.*, p. 324.

11. Lebbon, *op. cit.*, pp. 353, 363.

12. Greg Keyes, *Godzilla: King of the Monsters* (New York: Titan Books, 2019), pp. 126–127.

13. Lovelock, *Ages of Gaia*, p. 136.

14. Eiseley, *Invisible Pyramid*, pp. 61, 64–65, 104.

15. Ward, *Medea Hypothesis*, p. 90.

16. Keyes, *op. cit.*, p. 124.

17. *Ibid.*, p. 89. Jonah's involvement in the terrorist plot, given his prior history with illegally obtaining Titan DNA, initially confuses the Monarch team because his need for an ORCA device or reliance on Emma on how to utilize it isn't understood. After all, with most Titans in stasis, he could acquire his biological samples without the need for awakening or subsequently controlling the monsters. Emma's and Jonah's ulterior motive is far more devious than mere DNA-trafficking. (Also see Keyes, pp. 62, 64, 68–69 for clarification of this point.) In fact, devising the plot of culling the human population and revivifying Titans so that man could gradually learn once again to co-exist peacefully and in harmony with the monsters (as reflected in ancient records) was originally Emma's idea, not Jonah's (p. 134). But in the end she realizes that "Monster Zero isn't using the Titans to restore the planet—he's using them to destroy it. This isn't coexistence. It's extinction." Jonah expresses surprise for Monster Zero's actions, yet counters, "we already opened Pandora's Box. There's no closing it now. ... Humanity *is* a disease and the fewer of them there are the better it is for me" (Keyes, pp. 199–200).

18. For more on Rodan's rank as a "genuine" pterosaur dino-monster, see Chapter Fifteen in Debus's *Dinosaur Memories II, op. cit.*, pp. 149–157.

19. Keyes, *op. cit.*, p. 95.

20. McGhee, *Carboniferous Giants and Mass Extinction* (New York: Columbia University Press, 2018), p. 170. Could we suffocate if planetary oxygen diminishes? Well, atmospheric oxygen absolute levels are not sacrosanct, although regulated by Gaia. They've fluctuated before over the past 500 million years, roughly in the range of from 12% to 35% (while current atmospheric level is 21% at

sea level). Mass extinction events of the past and evolutionary innovation have certainly been tied to changes in atmospheric oxygen. Volcanic burning of then existing and extensive coal deposits would have exacerbated global warming, as well as further depress atmospheric oxygen levels. See Ward's *Out of Thin Air* (Washington, DC: Joseph Henry Press, 2006), and McGhee's *Carboniferous Giants*. Also see p. 442 in *The Earth System*, 3rd ed. (Pearson India Education Services, 2016).

21. Keyes, *op. cit.*, p. 144.

22. *Ibid.*, p. 127.

23. *Ibid.*, p. 192.

24. *Ibid.*, pp. 192, 196.

25. *Ibid.*, p. 202.

26. *Ibid.*, p. 239.

27. Mothra's battle with Ghidorah is described on pp. 273–286 of Keyes's novelization (*Ibid.*). Mothra is usually identified as a female natural entity. Scholars have often associated Nature and ecology with the female spirit. See, for example, Carolyn Merchant's *The Death of Nature: Women, Ecology and the Scientific Revolution* (HarperSanFrancisco, 1989). In Mothra's case, the formula goes like this: Nature identifies with female spirit. But in Toho's canon, Mothra is a female deified personification. Therefore, in the realm of giant monsters, Mothra *is* a full embodiment, symbolic representation of Nature, transforming in time via metamorphosis.

28. Keyes, *op. cit.*, p. 285.

29. *Ibid.*

30. *Ibid.*, p. 181.

31. McGhee, *op. cit.*, p. 214. According to Richard Muller, who in the 1980s named the Sun's hypothetical companion star "Nemesis"—the star postulated to have triggered comet swarms toward the inner solar system, setting a course of the dinosaur-killing impact event 66 million years ago—he had considered using the god-name Shiva instead, but felt it couldn't be used because it was already in use for a high-energy laser at the Lawrence Livermore National Laboratory. Nemesis was "the Greek goddess whose job it was to make sure that no earthly creature (e.g. the dinosaurs) challenged the dominance of the gods," struck down by a heavenly thunder bolt.

Muller, *Nemesis: The Death Star* (New York: Weidenfeld & Nicolson, 1988), p. 114.

32. Keyes, *op. cit.*, 2019, makes interesting, foreboding ties between the Permian extinction, our current Anthropocene "Age of Man" epoch, and science fictional planetary destruction wreaked by Ghidorah—the creature Emma resurrected using the ORCA (pp. 198–199). She horrifyingly realizes in the wake of Ghidorah's cataclysmic wave that Earth is facing a rapid mass extinction event, unwitnessed since the terminal Permian period.

33. This ensuing discussion is a bit convoluted and philosophical, and may stray a bit beyond the scope of filmic dino-daikaiju, but remains relevant at this juncture toward discussion of Gaian science. In his 2009 book *The Medea Hypothesis, op. cit.*, Ward states: "Life on Earth was not capable of predicting asteroid strikes, or preparing for them in any way. Hence these events had a Gaia-neutral standing. ... But what of ... others? Their cause seems to have been similar and in fact was life itself—at least, the effects of some species of life ... microbes gone wild" (pp. 81–82). However, even an extraterrestrial root cause such as an asteroid impact, such as the one known to have happened 66 million years ago—or that of a super-magmatic, carbon dioxide-releasing plume—may also still have Medean consequences. So, is the impact of a large mass extinction-causing asteroid or comet—or even a fictional/metaphorical Ghidorah "falling from the stars" to Earth in prehistoric times—to be classified as "Gaia-neutral," therefore not entirely Medean? Well, there's some leeway here (and a bit of philosophy)! As Ward notes, even Gaian-neutral root causes may still ultimately have Medean repercussions. In the instance of the Late Cretaceous mass extinction, for example, "While it is true that the impact was extraterrestrial, it could be argued that the effects of life *magnified* the extent of the extinction. By the end of the Cretaceous there were planet-spanning forests. One effect of the impact was the ignition of continent-spanning forest fires. This produced an enormous amount of

ash and dark carbon soot that filled the atmosphere. This soot, *a product of life*, caused global cooling for some months after the impact, which seems to have played a significant role in kill mechanism. Again, a Medean effect." (my italics, p. 88) Other mass extinction events, not *initiated* by *extraterrestrial* factors, would analogously seem to have been Earth-triggered (i.e., lacking an extraterrestrial root cause, therefore Medean), resulting in global greenhouse warming and then proliferation of (ordinarily) deep sea microbes toward deoxygenated oceanic surface waters in the throes of eutrophication—microbials that rely on sulfur as a nutrient and lethally emit poisonous hydrogen sulfide gas to the atmosphere. Their presence and pattern in promoting mass extinctions, such as during the Late Permian and other geological stages, is apparent in the rock record, in the form of black shale deposits found in multiple geological horizons and a chemical biomarker (isorenieratane) "characteristic of the purple and green sulfur photosynthesizing bacteria, forms that can live only in seas shallow enough for light to penetrate that are also low in oxygen and high in hydrogen sulfide concentrations." Ward, *Under a Green Sky*, p. 126. One driver for Ward's recurrent mass extinction scenario would be what has been referred to in the literature as the formation of a toxic, anoxic Canfield Ocean. (Ward, *op. cit.*, 2007, p. 125)

34. Graham Edwards, "Ancient Gods," *Cinefex 165*, no. 165 (June 2019), p. 57.

35. Keyes, *op. cit.*, pp. 301–302. It would seem that Ghidorah isn't finished yet … in 2019's film (as we shall see in Chapter Nine)!

36. Eiseley, *op. cit.*, p.115.

Chapter Seven

1. Don Glut quoted in Paul A. Woods, ed., *King Kong Cometh!* (Plexus Publishing, Ltd., 2005), p. 67.

2. Steve Vertlieb quoted in Woods, *op. cit.*, p. 84; As in the case of real dinosaurs viewed as scientific or political "objects," filmic dino-monsters, as well, may acquire an evolving stream of historically conflicting and variable pop-cultural meanings. Nieuwland, Ilja. *American Dinosaur Abroad: A Cultural History* (Pittsburgh, PA: Pittsburgh University Press, 2019), pp. 255, 258.

3. *Ibid.*, p. 175.

4. *Ibid.*, p. 190.

5. Pierce cited in J. D. Lees and Marc Cerasini, *The Official Godzilla Compendium* (New York: Random House, 1998), p. 17.

6. Steve Ryfle, *Japan's Favorite Mon-Star: The Unauthorized Biography of "The Big G"* (ECW Press, 1998), p. 14.

7. *Ibid.*, p. 43.

8. Ed Godziszewski, "The Making of Godzilla," in *Japanese Giants: The Magazine of Japanese Science Fiction and Fantasy*, no. 10 (Sept. 2004), p. 42.

9. Ryfle, *op. cit.*, p. 37.

10. *Ibid.*, pp. 56–58.

11. *Ibid.*, pp. 22–23.

12. *Ibid.*, p. 22.

13. Godziszewski, *op. cit.*, p. 37.

14. Pierce cited in J. D. Lees and Marc Cerasini, *op. cit.*

15. Mary Roach. *Gulp* (New York: W. W. Norton, 2013), p. 230.

16. Debus, *Dinosaur Memories II: Dino-daukaiju & Paleoimagery* (Createspace, 2017), pp. 347–354.

17. Movie review cited in Steve Archer, *Willis O'Brien: Special Effects Genius* (McFarland & Company, Inc., Publishers, 1993), p. 78.

18. Stephen A. Czerkas, *Major Herbert M. Dawley: An Artist's Life—Dinosaurs, Movies, Show-Biz, & Pierce-Arrow Automobiles* (The Dinosaur Museum, 2016), p. 299.

19. Godziszewski, *op. cit.*, p. 7.

20. *Ibid.*, pp. 10–11.

21. *Ibid.*, pp. 11–13, 23–30.

22. *Ibid.*, p. 29.

23. Archer, *op. cit.*, pp. 80–83.

24. See detailed discussion in Ryfle's *Japan's Favorite Mon-Star*, for example.

25. For example, *Famous Monsters of Filmland*, no. 35, October 1965 (Warren Publishing Co.), "Godzilla: King of the Creatures!" (attributed to Forrest J Ackerman), pp. 46–56.

26. Ryfle, *op. cit.*, p. 82.

27. William Tsutsui, *Godzilla on My Mind: Fifty Years of the King of Monsters* (Palgrave/ Macmillan, 2004), p. 100.

28. *Ibid.*, p. 101.

29. Ryfle, *op. cit.*, p. 83.

30. John LeMay, *The Lost Films: The Big Book of Japanese Giant Monster Movies* (Bicep Books, 2018), p. 89.

31. Donald F. Glut, *Classic Movie Monsters* (The Scarecrow Press, Inc., 1978), pp. 368, 371.

32. Walter Wager, *My Side by King Kong: As Told to Walter Wager* (New York: Simon & Schuster, 1976), p. 117.

Chapter Eight

1. J. I. Baker, *Life: Godzilla the King of Monsters* (New York: Life Books, 2019), p. 49.

2. *Ibid.*

3. *Ibid.*, p. 44

4. James Rollins, *The 6th Extinction: A Sigma Force Novel* (New York: William Morrow, 2014); James Rollins, *The Demon Crown: A Sigma Force Novel* (New York: William Morrow, 2017).

5. Nineteenth-century biologist Charles Darwin—an author of the theory of evolution via natural selection of species and who is a shadow character in Rollins's *6th Extinction*—lives in dread of the forbidding Antarctic shore where awful, DNA-alternate creatures must have evolved.

6. Rollins, *op. cit.*, p. 250.

7. *Ibid.*, p. 286.

8. For more on these book titles, see Debus's *Dinosaurs in Fantastic Fiction.* For more on the lost Spider Pit movie sequence, see "Mystery of the Spider-Men" in *Prehistoric Monster Mash: Science Fictional Dinosaurs. Fossil Phenoms, Paleo-pioneers, Godzilla & Other Kaiju-saurs* (Kindle/Amazon, 2019), pp. 86–96, by Allen A. Debus.

9. Rollins, *op. cit.*, pp. 265, 274.

10. James Rollins, *Subterranean* (New York: Avon Books, 1999). As in the case of Legendary's Skull Crawler threat witnessed in 2017's *King Kong*, in his writings Rollins explores the time-honored theme that "what stirs within the planet" is forbidden and menacing to man. Another more recent example of this idea was portrayed in a Netflix film, *The Silence* (directed by John R. Leonetti, 2019), in which flocks of blind, 5-foot-wingspan pterosaurian monsters (referred to as "Vesps" in the film) fly from an ancient cave penetrated by spelunkers. Whereas their kind has been concealed underground for millions of years (presumably since the Mesozoic when they evolved an adaptation for survival in utter cave darkness), since unleashed upon civilization, now they're deadly to humanity.

11. Rollins, *6th Extinction*, p. 214.

12. Rollins, *Ibid.*, p. 423.

13. Alida M. Bailleul, Wenxia Zheng, John R. Horner, *et. al.* "Evidence of proteins, chromosomes and chemical markers of DNA in exceptionally preserved dinosaur cartilage," *National Science Review*, Jan. 12, 2020 (published online); print copy (April 2020), vol. 7, no. 4, pp. 815–822; quote from p. 815.

14. Darlene Geis, *Dinosaurs and Other Prehistoric Animals* (New York: Grosset & Diunlap, Publishers, 1959), with pictures by R. F. Peterson, including those shown in *King Kong vs. Godzilla.*

15. J. D. Lees clarified where this rumor of an alleged double ending originated in issue 128 of *G-Fan* magazine, vol. 1 (Summer 2020), p. 57. The misleading suggestion first appeared in issue no. 7 of Ackerman's magazine, *Spacemen* (1963). Ackerman had stated, "Does King Kong best his saurian adversary or does Godzilla prevail over the mammoth ape? ... *Spacemen* magazine 'lets you in on a secret: 2 endings have been filmed & if you see King Kong or Godzilla in Japan, Hong Kong or some Oriental sector of the world, Godzilla wins! On the other hand, in the USA & England, for instance, Kong wins!' However, as Lees corrects, 'The erroneous passage in *Spacemen* magazine started a rumor that became 'common knowledge' and has persisted for more than half a century,'" perpetuated initially within pages of *Famous Monsters of Filmland* issues nos. 51 (August 1968) and later in 114 (March 1975). According to Svengoolie—Rich Koz's Chicagoland horror film television host, as broadcast on MeTV on Nov. 20, 2021— in the 1962 original Japanese movie, Godzilla's roar can be heard at the conclusion, while Kong swims off into the distance.

Chapter Nine

1. Alan Colosi. *KKXG: King Kong vs. Gigantosaurus (First Edition)—The Adventures of Yuriko Kumage During the Greatest War on Earth* (Shado Control, 2014/2016). Oddly, Colosi twice (e.g., p. 250) suggests his Gigantosaurus is also a "non-living" reptilian, which recalls Toho's original Godzilla in 1954's *Gojira*, a disfigured dino-monster that arguably could be perceived as dead-zombified, following its exposure to radiation from a hydrogen bomb explosion.

2. *Ibid.*, p. 6; Also, on p. 149, a character states, "It looks like a new evolution on the Earth ... Are we humans really the top of life and intelligence?"

3. *Ibid.*, p. 108.

4. There is also an enormous tyrannosaur living on East Island that stomps through a setting prior to a quick exit.

5. As stated on the back cover of Colosi's book.

6. I won't spoil what Kong's particular incentive happens to be here.

7. Colosi, p. 249.

8. *Ibid.*, p. 254. One "likely" theory for the monsters' disappearances is that "Meltdown inside Gigantosaurus got both of them..." (p. 258)

9. Story by Terry Rossio and Michael Dougherty and Zach Shields. In part due to the coronavirus pandemic, this movie had a rocky release date schedule. Originally the film had an intended release date of November 2019, which was pushed back into spring 2020, then bucked back a year to May 2021, but then moved up to March 2021. Most who saw the film probably watched it on HBO Max (as I did), instead of in a movie theater.

10. Russ Fischer. "Rampage! Godzilla vs. Kong Comes Out Swinging," *Fangoria*, no. 359, vol. 2, no. 11(April 2021), pp. 36, 39.

11. Wingard cited in *Ibid.*, p. 38.

12. *Ibid.*

13. Fischer, "Rampage," p. 38.

14. Hollow Earth theory has become a traditional staple, especially, of American science fiction. For example, most English and American readers who read Jules Verne's *Voyage au centre de la Terre* (1864, 1867) have read a translated version—usually an early 1870s editions retitled *A Journey to the Center of the Earth* (although there is a more recent English translation by William Butcher). This fascinating novel is of course the quintessential tale of intrepid explorers descending into the primeval bowels of our planet (although they never do reach Earth's center iron-nickel core). Indeed, the appealing concept of a hollow planet, and its prehistoric denizens, was perpetuated in another fantastic novel by written Edgar Rice Burroughs, *At the Earth's Core* (1914). Since then, other stories have been written about adventures *into* our planet, a theme also more famously projected over the decades by the movie industry. More recently, however, Lin Carter's novel *Journey to the Underground World* (1979) introduced readers to Earth's inner world of Zanthodon. Meanwhile, in Marc Cerasini's novel *Godzilla At World's End* (1998), giant monsters rove about below ground, surfacing to menace civilization until our fate is decided by an apocalyptical showdown on Antarctica between Godzilla and antagonist kaiju Ghidorah, Gigan and Hedorah. Connections between Hollow Earth theory and our planetary poles are most often dated back to American army officer and amateur scientist John Cleves Symmes, who professed in 1818, "I declare that Earth is hollow and habitable within ... that it is open at the poles." For more on Symmes, early scientific impressions, and reviews of several classic hollow Earth stories, see David Standish's *Hollow Earth: The Long and Curious History of Imagining Strange Lands, Fantastical Creatures, Advanced Civilizations, and Marvelous Machines Below the Earth's Surface* (Cambridge, MA: Da Capo Press, 2006). Symmes's quote is borrowed from p. 40 in this reference. Also, further extending the "holes at the poles theme," see the collection of stories reprinted in *The Antarktos Cycle: Horror and Wonder at the Ends of the Earth*, edited by Robert M. Price, including H. P. Lovecraft's *At the Mountains of Madness* and other chilling tales (Canada: Chaosium Inc., 1999). I have also written further about the sci-fi theme as it pertains to the prehistoric venue: see my McFarland

books, *Dinosaurs in Fantastic Fiction* and *Prehistoric Monsters* (listed in the Bibliography).

15. Keyes, *Godzilla vs. Kong*, pp. 61, 104–105.

16. *Ibid.*, pp. 59–61, 70–71.

17. In Keyes's 2021 novel, which at times seems *key*(-es) to understanding background in the film, we read on p. 175, "Ilene believed the Iwi and perhaps other humans had gone to war with Kong in the past, and there was evidence that Godzilla, too, might have had human followers."

18. Keyes, *Godzilla vs. Kong*, pp. 181–182.

19. *Ibid.*, p. 238.

20. *Ibid.*, pp. 189–190.

21. *Ibid.*, p. 117.

22. *Ibid.*, p. 133. Also, Ren Serizawa's perspectives on the Titans in the grand scheme of things is intriguing; Keyes offers key insights here as well on his dark persona in the 2021 novelization. "'Would you be proud of me, Father?' he wondered. Ichiro Serizawa had never understood the true potential of the beasts he spent his life studying. ... Humanity had always been beset by animals stronger, more deadly than its feeble primate members ... but risen above all of them. ... Bacteria and viruses were still worthy enemies, but for the most part the worst infectious diseases had been eradicated, and many of these organisms had been repurposed for genetic engineering. All this done by the physically weakest of all the great apes. ... The Titans, for all their size and power, they were just more of the same. The only question was whether they would be driven into extinction or repurposed for human ends. They were not gods; they were not worthy of worship—or of sacrifice. They were animals to be mastered, nothing more. His father ... had famously said, 'Let them fight.' ... only a man who did not care about human beings could say such a thing; only such a man could brush aside the untold casualties that 'letting them fight' always led to." Hubris in the extreme! (pp. 231–232)

23. *Ibid.*, p. 118.

24. *Ibid.*, p. 205.

25. *Ibid.*, pp. 264–265.

26. *Ibid.*, pp. 288–289.

27. *Ibid.*, p. 284.

28. *Ibid.*, p. 256.

29. *Ibid.*, p. 299.

Chapter Ten

1. Bill Bussone, "Kaiju Invasion," *G-Fan*, vol. 1, no. 127 (Spring 2020), p. 43.

2. Max Borenstein and Greg Borenstein, *Godzilla: Awakening* (Warner Bros. Entertainment, Inc., with Legendary Comics, 2014), illustrated by Eric Battle, Yvel Guichet, and Lee Loughridge, with cover art by Arthur Adams. (Note there is no page numbering within this short graphic novel.)

3. Arvid Nelson, *Godzilla: Aftershock* (Legendary Comics, The Official Prequel Graphic Novel to *Godzilla: King of the Monsters*, 2019), illustrated by Drew Edward Johnson. (Note there is no page numbering within this short graphic novel.)

4. This summary is borrowed from the back of the book, cited in note 3.

5. H. P. Lovecraft. "Dagon," a forward-projecting short story for the Chthulu mythos originally published in Nov. 1919 in *The Vagrant* no. 11, and reprinted in *100 Creepy Little Creature Stories*, selected by Robert Weinberg, Stefan Dziemianowicz, & Martin H. Greenberg. (New York: Barnes & Noble, 1994), pp. 86–90.

6. As Dalton Cooper aptly stated, recognizing Gaian implications, "the anime trilogy displays a *vengeful planet* attempting to destroy humanity through Godzilla." (my italics) Cooper, "Godzilla fan survey results," *G-Fan*, vol. 1, no. 131 (Spring 2021), p. 39. Cooper is writing a scholarly thesis on Godzilla for his master's degree.

Chapter Eleven

1. Ilja Nieuwland, *American Dinosaur: A Cultural History of Carnegie's Dinosaur Abroad* (Pittsburgh: University of Pittsburgh Press, 2019), pp. 26–27.

2. Edwin H. Colbert, *Men and Dinosaurs: The Search in Field and Laboratory* (New York: E. P. Dutton & Company, Inc., 1968), pp. 55–66.

3. Adrian J. Desmond, *The Hot-Blooded Dinosaurs: A Revolution in Palaeontology* (1975, New York: Warner Books ed., 1977), pp. 28–31.

4. Nieuwland, pp. 26–27.

5. Allen A. Debus (with Diane Debus), "Why Not the Mammals?" Chapter Five in *Dinosaur Memories: Dino-trekking for Beasts of Thunder, Fantastic Saurians, 'Paleo-people,' 'Dinosaurabilia,' and other 'Prehistoria'* (Lincoln: Authors Choice Press, 2002), pp. 85–110; Allen A. Debus, "Triumph of the Leapin' Lizards," Chapter Six in *Prehistoric Monsters: The Real and Imagined Creatures of the Past That We Love to Fear* (Jefferson, NC: McFarland & Company, Inc., Publishers, 2010), pp. 108–158.

6. I first encountered the term "cryptofiction" in an anomalously published introduction to a reprint of Henry Francis's 1908 short story, "The Last Haunt of the Dinosaur," (Read Books Ltd., 2013, A Cryptofiction Classic), title and p. 1.

7. Camille Flammarion, *Le Monde Avant La Creation De L'Homme* (Paris: C. Marpon et. e. Flammarion, 1886), p. 21; Allen A. Debus, *Prehistoric Monsters*, pp. 133–135; Allen A. Debus, *Dinosaur Memories II: Dino-Daikaiju & Paleoimagery* (Amazon Createspace, 2017), pp. 228–229, 231–233, 389. Surely through these early "sauropod" visuals, *Brontosaurus* must have seemed as Frankensteinic as Mantell's *Iguanodon*. For more on these voyeuristic dino-monsters, see Ulrich Merkl's 2015 book *Dinomania: The Lost Art of Winsor McCay, the Secret Origins of King Kong, and the Urge to Destroy New York* (Seattle: Fantagraphic Books), pp. 66, 69, 92. In one scene in 1925's *The Lost World*, Prof. Challenger's brontosaur-on-the-loose becomes the "peeping" dinosaur, its head smashing through an upper-story apartment window.

8. Nieuwland, *American Dinosaur*, p. 33.

9. While for many years *Brontosaurus* was considered a junior synonym to *Apatosaurus*, by 2016 the former is now a valid genus per Gregory S. Paul's book *The Princeton Field Guide to Dinosaurs*, 2nd ed. (Princeton, NJ: Princeton University Press, 2016), pp. 215–217. Also see paleontologist Matthew Mossbrucker interviewed by Tony Campagna in *Prehistoric Times* no. 132, Winter 2020, p. 33 for more on the reclassification of *Brontosaurus*.

10. Debus, *Prehistoric Monsters*, pp. 134–135, 141; Merkl, *Dinomania*, p. 69.

11. William D. Matthew, "The Largest Known Dinosaur. A Huge Extinct Reptile from German East Africa, the Largest Known Quadruped," *Scientific American*, vol. 111, no. 22 (Nov. 28, 1914), pp. 446–447.

12. For instance, see Francis's 1908 cryptofiction story, *op. cit.*; Edgar Rice Burroughs's *The People That Time Forgot* (1918, New York: Ace Books, 1963), pp. 45–46; Frank Saville's 1901 novel, *Beyond the Great South Wall* (New York: Grosset & Dunlap); Vladimir Obruchev's *Plutonia* (1924, Moscow: Raduga Publishers, 1988, translated by Fainna Solasko); and Lin Carter's *Zanthodon* (New York: Daw Books, 1980). H. G. Wells included a reference to a skeletal, museum *Brontosaurus* in his 1895 novella "The Time Machine." Harry M. Geduld, *The Definitive Time Machine: A Critical Edition of H. G. Wells's Scientific Romance* (With Introduction and Notes by Harry M. Geduld), Indiana University Press, 1987. Also see Allen A. Debus, "Wells versus Verne: Particulars in Paleo-Fiction," Chapter Thirty-four in *Prehistoric Monster Mash: Science Fictional Dinosaurs, Fossil Phenoms, Paleo-pioneers, Godzilla & Other Kaiju-saurs* (Amazon/Kindle, 2019), pp. 388–393. In two of John LeMay's books, he recounts several historical newspaper and magazine accounts from the late 19th and early 20th centuries, offering printed allegedly true, firsthand details concerning encounters between man and living dinosaurs. One interesting entry involved the Duke of Westminster, Hugh Grosvenor's travels to the relatively unknown and desolate Yukon Territory in search of a gigantic 70-foot-long living "Keratosaurus" (sp.). The story mainly played out in accounts published between 1907 and 1908, yet extending off and on well into the later 1920s. Another similar incident happened in 1916 with a report of an alleged Tyrannosaur attack on Eskimos near the North Pole. While details of the deadly and uncanny "real"

encounter were reported as factual news in the *Ogden Daily Standard*, LeMay suggests this elaborate tale was instead clearly "an obvious work of fiction." See LeMay's *Cowboys & Saurians: Ice Age* (Roswell, NM: Bicep Books, 2020), Chapters One and Twenty, pp. 15–65 and 230–244, respectively.

13. As Merkl *op. cit.* suggests in his lavishly presented volume, McCay was pondering the inadvertent, harmless yet very real destruction of metropolitan centers in his unfinished cartoon of 1919–1921, *Gertie on Tour*, and his unpublished comic strip, "Dino," begun and roughed out by the early 1930s (i.e., 1934). Some scenes planned for the former, but never adapted into actual footage show Gertie using the Brooklyn Bridge as a trampoline, yet with the words "She means no harm" handwritten at the top of the sketch (Merkl, p. 105), and also uprooting the Washington Monument (Merkl, p. 105). Meanwhile his super-large "Dino" was shown in several panels smashing his way through New York City. Despite such visuals taken at first glance, reckless destruction wreaked by McCay's imagined city-smashing sauropods was unintentional, not driven apocalyptically. As Merkl suggests, the sequence of Gertie's c. 1921 *On Tour* scenes may be considered precursors for McCay's later day "Dino" strip created during 1933–1934, which never made it to print. Although resembling Gertie, from surviving sketches and Merkl's own inferred efforts, recently relying on Roger Langridge to reconstruct missing cartoon panels, McCay's "Dino," "lost art" dino-monster sauropod character—an evidently gentle creature yet utterly misplaced in modernity, seemed inadvertently bent on rampage in metropolitan areas. As Merkl opined, the Dino strip would have been a "masterpiece very nearly (making) its debut (i.e. in the fall of 1934): Winsor McCay's Dino, the odyssey of a dinosaur who awakens from a long, deep sleep and starts getting into trouble with human civilization. But it was not to be" (Merkl, p. 15) See Debus, "When Brontosaurus Was Monstrous," Chapter Twenty-six in *Prehistoric Monster Mash, op. cit.*, pp. 291–309. McCay is also credited

with producing a precursor to the giant-monster-attacking-metropolitan-center sci-fi theme in a ~12-minute feature animated film titled *The Pet* (1921), available on YouTube. However, this strange monster conceived over three decades before the Rhedosaurus and Godzilla *isn't* a dino-monster, but instead a curious *daikaiju*-like "puppy" that rapidly grows to immense sizes before succumbing to a squadron of military biplanes and dirigibles.

14. Samuel Hopkins Adams's (1871–1958) *The Flying Death* (serialized 1903, reprinted New York: McClure, 1908) involved a wayward, murderous *Pteranodon*. Jules Lermina's (1839–1915) *Panic in Paris* (1910, 1913 as *L'Effrayante aventure:* Encino, CA: Black Coat Press, 2009, adapted by Brian Stableford) drafted *Megatherium* and both an *Iguanodon* and *Brontosaurus* among other menacing prehistoria, for a deadly prehistoric invasion from within the Earth underlying Paris. Although initially unimpeded by the military, as Lermina stated of their eventual demise in a rather Wellsian way, "These unfrozen denizens of the Quaternary had only been reanimated by a false, temporary life. They bore ... the burden of their antiquity and their decrepitude—and, one by one, under the pressure of the ambient air, beneath the spring sunshine, ill-adapted for life in an atmosphere hundreds of thousands of years younger than the one they had previously breathed, they were dying, being too old kin a world that was too new" (p. 191). A peculiar April Fools lark was printed in the 1906 *Chicago Sunday Tribune*, titled "Chicago Invaded by Hordes of Prehistoric Monsters Dealing Death and Destruction." For more on this item, see Merkl, *op. cit.*, pp. 82–84, and Debus, "Chicago Attacked by a Swarm of Giant Dinosaurs Storming From Canada," Chapter Five in *Dinosaur Memories II: Pop-cultural Reflections on Dino-Daikaiju and Paleoimagery* (Amazon Createspace, 2017), pp. 56–61. In a climactic battle described in the 1906 newspaper account, a *Diplodocus* battles a *Tyrannosaurus* to the death on the Montgomery Ward building tower on Michigan Avenue in Chicago. Meanwhile the military engages the invading hordes of dino-monsters. None of

these fictional occurrences were in any sense "paleo-apocalyptical," and, in the case of the *Chicago Sunday Tribune* story, were merely an entertaining bit of good-natured tom-foolery. For more on Hopkins's novel, see Debus, *Prehistoric Monster Mash, op. cit.,* pp. 137–142. Such prehistoric invaders were dispatched in relative short order.

15. Stephen J. Gould, "Dinomania," in *Dinosaur in a Haystack: Reflections in Natural History* (New York: Harmony Books, 1995), pp. 221–237; Mitchell, *Last Dinosaur Book,* pp. 9–14. Ultimately, Gould dismissed this conclusion founded in "archetypal fascination," instead suggesting that the true answer lay in blatant consumer commercialization (p. 224).

16. Mitchell, *Last Dinosaur Book,* pp. 12–13.

17. Debus, *Dinosaur Memories II,* pp. 188–201.

18. Nieuwland, *American Dinosaur,* pp. 42–43, 97. The hugeness zeitgeist factor is a key theme throughout Merkl's 2015 book.

19. Allen A. Debus, "Big, Fierce, Extinct ... Radioactive: Godzilla's Essential Formula," *G-Fan,* no. 115, Spring 2017, pp. 68–75; Nieuwland, *op. cit.,* p. 254.

20. Allen A. Debus. "All-time Greatest Prehistoric Monsters of Science Fiction: A Panel Experience," *G-Fan,* vol. 1, no. 122 (Winter 2018), pp. 32–36.

21. Nieuwland, *American Dinosaur,* p. 73.

22. *Ibid.,* p. 100.

23. *Ibid.,* p. 84.

24. There were at least two (other) not so friendly, if not heated, competitions, controversies leading toward lawsuits involving sauropods during this early period. First, the aforementioned Winsor McCay's work proved so successful as to invite imitation from New York animator John R. Bray (1869–1932), who according to Donald Glut made his own "Gertie the Dinosaur" cartoon. However, "this ersatz Gertie lacked the original's fluidity of movement and amiable personality." Glut, *The Dinosaur Scrapbook* (Secaucus, NJ: Citadel Press), pp. 168, 170. Merkl in *Dinomania* refers to Bray's 1915 (~7-minute-long) copycat production (also available on YouTube) as a "ripoff."

(Merkl, p. 281 n. 133) McCay sued Bray and prevailed, but the latter was soon at it again with a Gertie-like *Diplodocus* cartoon character. Also in the arena of life-sized mechanical/animatronic brontosaurs, another legal case ensued during the early 1930s over patent rights, as in the matter of George H. Messmore's and Joseph Damon's 49-foot-long "Dinah," versus Earl Carroll's "Sarah," further outlined in my *Dinosaur Memories II,* in a subsection of Chapter Six, "Giant Dino-Monsters & 'Kong' Invade Chicagoland," pp. 65–66.

25. Stephen Czerkas, "O'Brien vs. Dawley: The First Great Rivalry in Visual Effects," (*Cinefex* 138, July 2014), pp. 13–27; Stephen A. Czerkas, *Major Herbert M. Dawley: An Artist's Life—Dinosaurs, Movies, Show-Biz, & Pierce-Arrow Automobiles* (Blanding, UT: The Dinosaur Museum, 2016); Allen A. Debus, *Prehistoric Monster Mash,* pp. 291–309. How sauropods actually moved and lived has remained a subject of scientific controversy for over a century. See Adrian J. Desmond's *The Hot-Blooded Dinosaurs: A Revolution in Paleontology* (1975, New York: Warner Books, 1977 ed.) for example.

26. In the matter of Kong, Merkl, *op. cit.,* states, "Surely there are too many parallels here for this to be a coincidence" (p. 205). Also, in lamenting the "lowly status of the comic strip," (pp. 191, 251), he generally "cannot prove it" (p. 129). See Ulrich Merkl's *Dinomania,* 2015.

27. Winsor McCay generally seems to have been fascinated with relative sizes of *giant* creatures which he featured in cartoons, human forms, as well, ambling about through New York City. Merkl (pp. 129, 191, 205, 251) also suggests that McCay's sauropod dinosaur at loose-in-the-city-themed comics may have influenced others to take up the calling later in prehistoric dino-monster attack films. While he makes a plausible and interesting yet circumstantial case for McCay having possibly had some sort of subliminal bearing on Merian C. Cooper's *King Kong,* such evidence is either lacking or, at best, derivative for inspiring Toho's *Gojira.* When it comes to Godzilla, Winsor McCay's

influence may be glimpsed, too, but vastly diluted genetically like a great, great paleo-parent. One must be cautious to avoid the fallacious argument that simply because something came before does not necessarily mean it caused something to result later that might be similar or seemingly derivative in nature. (The Latin term for this is *post hoc ergo propter hoc.*) See Debus, Chapter Eighteen in *Dinosaur Memories II, op. cit.,* pp. 188–201. Mention should also be made here of the first life-sized sauropod statue restoration, a *Diplodocus* constructed by Joseph Pallenberg (1882–1946) for Carl Hagenbeck's "Tiergarten" Zoological Gardens in Hamburg, Germany. This statue was also shown in a film documentary, Max Fleischer's *Evolution* (Red Seal Pictures/Inkwell Studios, 1923, directed by M. Fleischer), although therein this dinosaur was referred to as *Brontosaurus.* Nieuwland claims that by circa 1910 *Diplodocus* was "perhaps the best-known dinosaur in the world" (p. 198). As well-known as *Diplodocus* had become by the 1910s, thanks to efforts of the Carnegie Museum of Natural History, arguably *Brontosaurus* was more representative of the clan. Throughout the 1910s and 1920s, sauropods were generally becoming characterized in public consciousness as small-brained, slow and dull-witted (Nieuwland, *op. cit.,* pp. 149, 255). If anything, gigantism was becoming synonymous with the obsolete or the inevitability of extinction. Despite their undisputed gigantism, and although several sauropods in popular culture were occasionally portrayed as fierce in persona, it was increasingly recognized that *other* dinosaurians were more worthy of such repute. Clearly Carnegie's "cosmopolitan" *Diplodocus* was losing luster, relative to *Brachiosaurus* and *Brontosaurus,* by circa 1910. Nieuwland, *op. cit.,* pp. 239–240.

28. Francis, "The Last Haunt of the Dinosaur."

29. Literature stories recruiting marauding, science fictional sauropods of the 1920s and beyond into the early postwar period include Willis Knapp Jones's "The Beast of Yungas" (Rural Publishing Company, 1924; reprinted in *100 Creepy Little Creature Stories,* New York: Barnes & Noble, 1994, pp. 44–51); Katherine Metcalf Roof's "A Million Years After" originally published in *Weird Tales* (Nov. 1930); John Hunter's serialized novella "Menace of the Monsters" (*Boys' Magazine,* 1933/34); and a pair of novels by John Russell Fearn (aka Vargo Statten), *A Thing of the Past* (1953; Wildside Press LLC, 2003) and *The Genial Dinosaur* (1953; Wildside Press LLC, 2012). The last three titles were written for younger to adolescent readers. For more on Roof's story, see Debus, *Dinosaurs Ever Evolving* (Jefferson, NC: McFarland & Company, Inc., Publishers, 2016), pp. 76–77. For more on Hunter's story, see Debus, *Prehistoric Monster Mash,* pp. 174–178. For more on Fearn's novels, see Debus's article, "Dino-Monster to the Rescue versus Aliens in 1953!" in *G-Fan* no. 107, Fall 2014, pp. 66–68.

30. Merkl, *Dinomania,* p. 134.

31. The three Knight paintings mostly evoking the savage theme of the time are his oft-reprinted depiction of two fighting Laelaps for an 1897 issue of *Century Magazine* (William Ballou, Nov. 1897, vol. LV, no. 1, pp. 10–23), an American Museum of Natural History tyrannosaur versus *Triceratops* portrayal of c. 1906, and the cited Chicago Natural History Museum's large late 1920s mural.

32. The familiar "dino-monsters run amuck in a city, fighting each other" theme was introduced over a century ago in literature. It would be another four decades, however, before the first pseudo-dinosaurian battle fought between a pairing of giant monsters (unleashed by the hydrogen bomb) in a major city (Osaka) would be dramatically staged on film, as witnessed in Toho's *Godzilla Raids Again* (1955). "Gigantis" triggered an onslaught of giant paleo-monster rallies such as *King Kong vs. Godzilla* and *Ghidrah, the Three-headed Monster.* Today, in the modern cultish era of battling *daikaiju,* this apocalyptical theme has proven to be incredibly popular, both in film and literature. And it's really taken off considerably further since the day of those "Dinosaurs Attack!" trading cards issued not so long ago (Topps, 1988), invoking a parallel universe

where humanity had not just one dinosaur to tangle with, but a ferocious saurian army attacking out of time. Merkl, *op. cit.*, suggests that a psychological urge to (symbolically) destroy New York City in comic strips (such as McCay's) or later in films (such as in *King Kong*, etc.) lay in underlying technophobic, Luddite tendencies: rejection of urban life and its pollution.

33. Debus, *Dinosaur Memories II*, pp. 188–201.

34. This is a theme prevalent throughout Spencer R. Weart's 1988 book *Nuclear Fear: A History of Images* (Cambridge, MA: Harvard University Press). Also see Richard Rhodes's *Dark Sun: The Making of the Hydrogen Bomb* (New York: Simon & Schuster, Touchstone ed., 1996).

35. H. G. Wells, *The World Set Free* (1914, Amazon ed. reprint, described further in Section 3). Apocalyptical science fiction may be said to have originated during the 19th century prior to Wells's novel. Of particular interest may be the genre of "last man" stories, of which Mary Shelley, author of *Frankenstein, or Modern Prometheus*, penned the first modern such version—1826's *The Last Man*, concerning the sole survivor of a total pandemic. (For more on Shelley's novel, see Debus, Chapter Thirty-five, "It's Alive!!," in *Prehistoric Monster Mash*, pp. 399–415.)

36. In Lermina's *Panic In Paris, op. cit.*, discovery of a powerfully energetic (i.e., radioactive) mineralogical substance "vrilium" facilitates an invasion of reanimated prehistoric monsters. (*Diplodocus* is mentioned in one passage.) For more on the association between atomic missiles and dinosaurs in *The Lost Continent*, see Debus, *Dinosaur Memories II* (2017), Chapter Four, "Revisiting the Lost Continent," pp. 46–50.

37. Peter J. Bowler, *A History of the Future: Prophets of Progress from H. G. Wells to Isaac Asimov* (Cambridge: University of Cambridge Press, 2017), pp. 151–152.

38. Ironically, to combat the dino-monster and its super-spreader germs, using a rifle, a marksman injects a good (thus benefiting mankind's immediate

problem) radioactive isotope into its bloodstream.

39. Michihiko Hachya, M.D., *Hiroshima Diary*, translated and edited by Warner Wells, M.D. (New York: Avon Publications, 1955); Oishi Matashichi, *The Day the Sun Rose in the West: Bikini, the Lucky Dragon, and I* (University of Hawai'i Press, 2011); Allen A. Debus, *Dinosaurs Ever Evolving: The Changing Face of Prehistoric Animals in Popular Culture* (Jefferson, NC: McFarland & Company, Inc., Publishers, 2016), Chapter Nine "Nuclear Dragon: Godzilla and the Cold War—1954," pp. 114–126.

40. For instance, see Isaac Asimov, *In Joy Still Felt: The Autobiography of Isaac Asimov, Vol. 2 1954–1978* (Garden City, NY: Doubleday and Company, Inc., 1980), pp. 9–10, where a 1954 paper "The Radioactivity of the Human Body" is discussed.

41. Spencer R. Weart, *The Rise of Nuclear Fear* (Cambridge, MA: Harvard University Press, 2012), p. 97.

42. Bill Warren, *Keep Watching the Skies—21st Century Edition* (Jefferson, NC: McFarland & Company, Inc., Publishers, 1982, 2010), Preface, p. 9.

43. Weart, *Nuclear Fear*, 1988, p. 325.

44. *Ibid.* Weart further states, "Many ... environmental writings likewise used images taken over directly from nuclear fears. ... John Wyndham's *Re-Birth* (1955) with industrial pollution replacing bombs as the cause of humanity's downfall and transmutation" (p. 325).

45. Weart coins the metaphoric term "human volcano" for our industrialized pollution of the natural environment in his 2003 book *The Discovery of Global Warming* (Cambridge, MA: Harvard University Press), p. 97. During the late 1940s—as suggested visually via Zallinger's *Age of Reptiles* mural—and into the 1950s, one popular theory for why dinosaurs became extinct in the Late Cretaceous was due to worldwide intensified volcanism, thereby contaminating Earth's ocean-atmosphere system, a theory that was resurrected during the 1980s as well. Thus, mankind's possible pending extinction could be likened to that of the non-avian dinosaurs.

46. I have endeavored to outline these periods extensively in my 2016 book

Dinosaurs Ever Evolving, op. cit., amending Chapter Two of *Dinosaur Sculpting: A Complete Guide, 2nd ed.*, A. Debus, D. Debus and B. Morales (Jefferson, NC: McFarland & Company, Inc., Publishers, 2013).

47. A general theme discussed in A. Debus, *Dinosaurs Ever Evolving.*

48. *Ibid.* A more recent wonderful example of the intelligent dinosaur-dinosauroid concept may be enjoyed in Heidi K. Richter's well-crafted 2016 novel *Red Feather* (heidirichter.com/redfeather). Here, species of flightless avian dinosaurs surviving on an island since the Mesozoic ultimately align themselves against plundering mankind. As one of the intelligent dinosaurians, known as Pteroraptors, explains: "Sanctuary ... so the island ... And the sanctuary Graniteclaw built. The two times we have called for sanctuary from extinction. The asteroid and then the people that moved onto this continent a few generations ago, killing everything as they went. ... One last try to do whatever we can to survive" (p. 279). Humans, like the asteroid of 66 million years ago, each in their own way threaten dinosaurs of past and present. Another flock of flightless, cunning dinosaurian-like, ten-foot-tall terror birds, *Titanis walleri*—survivors from the Pleistocene—hunt man hungrily in James Robert Smith's gripping 2006 novel *The Flock* (Waterville, ME: Five Star-Tekno Books). Upon first sight, one doomed character in the novel sighs, "A dinosaur ... Jurassic Park for real" (p. 89). They're possessed with "an intelligence that might match that of humans" (p. 323). Thus, reflecting the human condition, it would seem that in the minds of talented sci-fi writers, raw, unbalanced intelligence proves more menacing to man's survival (and Gaia) than wisdom.

49. Radiation-themed causal factors of potential doom in science fiction continued to haunt numerous television shows and movies—several involving or incorporating dino-monsters—through the late 20th century. Even in the new millennium, dangers of radioactivity—posed by metaphorical dino-monsters of doomsday—were not neglected. James Blish (1921–1975) ominously associated dinosaurians with dangerous radioactive materials in a pairing of novels, *A Case of Conscience* (New York: Ballantine Books, 1958) and *The Night Shapes* (New York: Ballantine Books, 1962). Nuclear threats evident in the 1961 film *Gorgo* (MGM, directed by Eugene Lourie) remained subdued to appease audiences, whereas Bruce Cassiday's aka—"Carson Bingham's" 1960 *Gorgo* novelization (Derby, CT: Monarch Books) from the movie script portended of man's coming apocalyptical plight (p. 73). By the turn of the new century, and into the new millennium, reverberating Godzilla's triumphant roar from half a century earlier, radiation remained at the core of several other authors' minds who associated dino-monsters with radiation, as we read in novels like Leigh Clark's *Carnivore* (Book Margins, Inc., 1997), and Matthew Dennion's *Atomic Rex* (Hobart, Tasmania: Severed Press, 2015). A pairing of movies accented by novelizations appeared in 1998, relating emergence of their featured dino-monsters with nuclear bomb tests or radioactive waste dumping—*Godzilla* (TriStar, directed by Roland Emmerich; Stephen Molstad's novelization printed by HarperPrism) and *Gargantua* (Fox Network, directed by Bradford May; Tor Books novelization by "K. Robert Andreassi," aka Keith Candido). For more on *Gargantua*, a dino-daikaiju relatively unknown today, see Chapter Thirteen in Debus, *Prehistoric Monster Mash.*

50. David Sakmyster and Rick Chester, *Jurassic Dead* (Hobart, Tasmania: Severed Press, 2014); Rick Chester and David Sakmyster, *Jurassic Dead 2: Z-volution* (Hobart, Tasmania: Severed Press, 2015); Rick Chester and David Sakmyster, *Jurassic Dead 3: CRTL-Z* (Hobart, Tasmania: Severed Press, 2015).

51. Sakmyster and Chester, *Jurassic Dead*, p. 145.

52. Michael Crichton wrote this compelling dialog for his novel character "chaotician" Ian Malcolm: "We are witnessing the end of the scientific era. Science, like other outmoded systems, is destroying itself. As it gains power, it proves itself incapable of handling the power. Because things are going very fast now. Fifty years ago, everyone was gaga

over the atomic bomb. That was power. No one could imagine anything more. Yet, a bare decade after the bomb, we begin to have genetic power. And genetic power is far more potent than atomic power. And it will be in everyone's hands. It will be in kits for backyard gardeners. Experiments for schoolchildren. Cheap labs for terrorists and dictators. And that will force will everyone to ask the same question—What should I do with my power?—which is the very question science says it cannot answer." (New York: Ballantine Books, p. 313)

53. H. G. Wells, *The Island of Dr. Moreau* (1896) in *The Complete Treasury of H. G. Wells* (New York: Avenel Books, 1978), pp. 67–157; Aldous Huxley, *Brave New World* (New York: Harper & Row, Publishers, 1932, 1978 ed.).

54. For more on this obscure title see Claudine Cohen's *The Fate of the Mammoth: Fossils, Myth, and History* (Chicago: University of Chicago Press, 2002, translation by William Rodarmor), pp. 13–14.

55. Another obscure, but eminently readable title in this vein is Robert Wells's *The Parasaurians* (New York: Berkley Medallion Books, 1969), discussed within Chapter Seventeen, "Dinosaurs in Fantastic Fiction—Extras" of my 2017 *Dinosaur Memories II*, pp. 162–187.

56. Jeremy Robinson, *Project Nemesis* (Breakneck Media, 2012), p. 297. Clearly, Nemesis is a daikaiju, but is it also a *dino*-daikaiju? As I've discussed in *Prehistoric Monster Mash, op. cit.*, pp. 267–269, the answer is resoundingly "no." However, based on its general appearance such as described within passages in Robinson's novels, further conveyed by artist Cheung Chung Tat's masterful cover artwork, and if such a kaiju monster appears pseudo-dinosaurian—generally resembling creatures from our Mesozoic age—then through its imagery and besides its cataclysmic actions the monster may also effectively conjure or signal thoughts of our impending extinction. One may then perhaps consider it a plausible pseudo-dino-daikaiju. Robinson had this to say about kaiju in an afterword note to his 2015 novel *Project Hyperion* (Breakneck Media, 2015): "When I started working on Nemesis

... my lofty goal was to create America's first real iconic kaiju. Sure, we have King Kong, but he's just a big ape and a snack for Godzilla, who Americans have adopted as their own. But the big green G will always be, at heart, Japan's kaiju. And now we have the kaiju in *Pacific Rim*, but they're hardly the main focus of the movie. There is *Cloverfield*, but can anyone even remember what that monster looks like? ... Nemesis ... started as the bestselling (not Godzilla) kaiju novel ever" (p. 267). Also see Robinson's *Project Maigo* (Breakneck Media, 2013), pp. 54, 243. For more on kaiju matters and definitions, see the Appendix.

57. Other novels in Robinson's 'Nemesis' series are: *Island 731* (New York: Thomas Dunne Books, St. Martin's Press, 2013); *Project Maigo, op. cit.*; *Project 731* (Breakneck Media, 2014); *Project Hyperion, op. cit.* And as in Shelley's foundational novel *Frankenstein*, Robinson clarifies who the real monsters are, with a twist: "Becoming a monster might be the only way the human race survives. ... It might be time to reassess what defines a monster, seeing past what's on the outside and evaluating what is on the inside." *Project Hyperion, op. cit.*, pp. 170–171. Furthermore, in Robinson's novels, American scientists are working for an agency named the Genetic Offense Directory, which has the revealing acronym "GOD." There is an intriguing extraterrestrial tie in Robinson's *Nemesis* series that I won't spoil here for readership. However, there are also interesting alien ties related to the Godzilla series of films (movie scripts) as well—that is, besides interstellar Ghidrah—that are worth referencing here, as researched by John LeMay. For these references, see A. Debus, Chapter Twenty-four, "Coming Dark Ages: Four Metaphorical 'Horsemen' of the Paleo-pocalypse," in *Prehistoric Monster Mash*, pp. 267–270. Al Gore wrote in 1992, "Some even imagine that genetic engineering will soon magnify our power to adapt even our physical form. We might decide to extend our dominion of nature into the human gene pool, to not just cure terrible diseases, but to take from God and nature the selection of genetic variety and robustness that gives our species its resilience

and aligns us with the natural rhythms in the web of life. Once again, we might dare to exercise godlike powers unaccompanied by godlike wisdom." Gore, *Earth in the Balance: Ecology and the Human Spirit* (New York: Houghton Mifflin Company, 1992), p. 240.

58. Bill Gates (*Time*, Jan. 15, 2018, p. 42), cited in A. Debus, *Prehistoric Monster Mash, op. cit.*, p. 261. Finally, in the Dinosaurs Attack! set of Topps 55-trading cards (1988) and through five comic books, mankind is imperiled, forced nearly to the brink of extinction, by a horde of prehistoric dino-monsters emerging from a mad "modern miracle" of futuristic science known as TimeScan. In the end—ironically, across a breach in time—it is man who exterminates dinosaur. (Dinosaurs Attack!, Feb. 2014, San Diego, CA, IDW Publishing, Topps Company, graphic novel by Gary Gerani, Herb Trimpe, Erl Norem, Flynt Henry; "Gary Gerani—Creator of the 1988 Dinosaurs Attack! Card Set," in Prehistoric Times, no. 136, Winter 2021, Gerani interviewed by Mike Fredericks, pp. 50-51). Archaeological, genetic and geological evidence indicates that endangered Paleo-Man's early, near-extinction actually has hung in the balance through several remote stages of the Pleistocene (The UnXplained, S2: E22, hosted by William Shatner, "Apocalypse When?" first broadcast on The History Channel, 11/26/21) from a number of causes, most recently during the Lower Dryas (12,900 years ago)—now recognized as resultant of a comet impact. One may distinguish the very recent Anthropocene's "6th extinction" of modernity with concomitant industrially caused global warming from that, resulting in extermination of mammalian megafauna, by the Holocene epoch's advent (12,000 to 11,500 years ago). (James Lawrence Powell, Deadly Voyager: The Ancient Comet Strike That Changed Earth and Human History, deadlyvoyager.net, 2020) Some philosophers might regard such extinctions of the past 13,000 years as "unnatural" due to extraterrestrial causal and human interventions. (Also see The Cycle of Cosmic Catastrophes: Flood, Fire, and Famine in the History of Civilization, by Richard Firestone, Allen West,

and Simon Warwick-Smith, Rochester, Vermont, Bear & Company, 2006).

Epilogue

1. Rhoads and McCorkle, *Japan's Green Monsters*, pp. 177–178.

2. Christopher Manes, *Green Rage: Radical Environmentalism and the Unmaking of Civilization* (Boston: Little, Brown and Company, 1990), Paul Ehrlich quoted from a 1987 interview on p. 40.

3. Rachel Carson, *Silent Spring* (Greenwich, CT: Fawcett, 1962), p. 168. For more on Carson and her concept of a balance in natural systems, adding to her influence on Lovelock, see Gribbin, *He Knew He Was Right*, pp. 104–110.

4. *Ibid.*, p. 169.

5. Senator Al Gore, *Earth in the Balance: Ecology and the Human Spirit* (New York: Houghton Mifflin Company, 1992), Chief Seattle quote printed on p. 259.

6. Carson, *Silent Spring*, pp. 218, 261.

7. Gore, *Earth in the Balance*, p. 91.

8. *Ibid.*, pp. 127–129. Thomas Robert Malthus (1766–1834) was a British economist who realized that during the Industrial Revolution rapidly increasing human populations would eventually outrun food supply, thus creating survival bottlenecks. His work was greatly influential toward the development of Charles Darwin's theory of evolution.

9. Mike Bogue, *Apocalypse Then: American and Japanese Atomic Cinema, 1951–1967* (Jefferson, NC: McFarland & Company, Inc., Publishers, 2018), p. 198.

10. *Ibid.*, pp. 199–200.

11. Rhoads and McCorkle, *Japan's Green Monsters*, p. 78; Peter J. Bowler mentions that during the 1960s, social and evolutionary biologists expressed a "real concern ... that without some system to control reproduction, any progress in food production would simply lead to further population expansion and the ultimate exhaustion of the planet's natural resources. ... (Authors) stressed that all efforts to use technology to tackle the problem of poverty had failed. ... (Paul) Ehrlich echoed ... that efforts to expand the food supply were resulting in the destruction of the planet. Environmentalism was a growing

concern (highlighted ... by Apollo 8's photographs of the Earth from space)." Bowler, *A History of the Future*, pp. 202–203. Christopher Manes also weighed in on the "population bomb" problem in his book *Green Rage*, writing, "with the population doubling every half century in a world already undergoing biological meltdown, it is difficult to conceive how human population can draw down in a sensible manner without the catastrophe Ehrlich foresaw" (p. 233).

12. Garrett, *Coming Plague*, Sir Burnet quoted on pp. 213–214.

13. Manes, *Green Rage*, Vajk quoted on p. 160.

14. Rhoads and McCorkle, *Japan's Green Monsters*, p. 77.

15. *Ibid.*, p. 78.

16. Gore, *op. cit.*, p. 92.

17. Rhoads and McCorkle, *Japan's Green Monsters*, p. 115.

18. *Ibid.*, pp. 116–117, 181.

19. *Ibid.*, p. 121.

20. *Ibid.*, p. 124.

21. *Ibid.*, pp. 122–123.

22. *Ibid.*, p. 117.

23. *Ibid.*, pp. 105–111.

24. *Ibid.*, p. 108.

25. *Ibid.*

26. *Ibid.*, p. 106.

27. *Ibid.*, p. 109.

28. *Ibid.*

29. *Ibid.*, pp. 110–111.

30. *Ibid.*, p. 136.

31. *Ibid.*, p. 130.

32. Tester quoted in Garrett, *Coming Plague*, p. 562.

33. Rhoads and McCorkle, *Japan's Green Monsters*, p. 149.

34. *Ibid.* What is natural? In the traditional man versus nature theme, given that man and all his machinations are natural, then isn't the conflict simply man versus himself?

35. *Ibid.*, p. 152. Expressing her concerns with the tampering of genetic systems in agricultural plants, Garrett, *op. cit.*, stated, "Genetic change in plant microbes was accelerating due to agricultural practices that exerted strong selection pressures on the microbes; the changing geography of plant growth due to international trading of plant seeds and breeding practices; and to the deliberate release of laboratory genetically

altered plant viruses that were intended to offer agricultural crops protection against pests. ... plant cells could swap genes with other viruses in the plant, producing active, pathogenic—*new*—viral species" (pp. 577–578). Also see Gore, *Earth in the Balance*, pp. 130–131.

36. Carson, *Silent Spring*, p. 161.

37. *Ibid.*, p. 233

38. *Ibid.*, p. 236.

39. *Ibid.*, p. 39.

40. *Ibid.*, p. 204.

41. *Ibid.*, p. 184.

42. *Ibid.*, p. 18.

43. *Ibid.*, p. 43.

44. *Ibid.*, p. 253.

45. Gore, *Earth in the Balance*, p.188.

46. *Ibid.*, p. 157.

47. *Ibid.*, p. 39.

48. *Ibid.*, p. 162.

49. *Ibid.*, pp. 50–55.

50. *Ibid.*, pp. 74, 79.

51. *Ibid.*, pp. 305–317.

52. *Ibid.*, pp. 131–138, 144, 240. Shades of Toho's bioengineered Biollante!

53. *Ibid.*, pp. 216–217. One of the leaders of Earth First!, Mike Roselle, a group espousing Deep Ecology, has said, "You hear about the death of nature and it's true, but nature will be able to reconstitute itself once the top of the food chain is lopped off—meaning us" (quoted in Gore, p. 217). Arne Naess coined the term "Deep Ecology" (Manes, *Green Rage*, pp. 139–140).

54. For a full outlay on Gore's Global Marshall Plan proposal, see Gore, *op. cit.*, pp. 295–360.

55. Manes, *Green Rage*, p. 31.

56. For more on America's first Earth Day, see Manes, *Ibid.*, pp. 45–51.

57. Foreman cited in Manes, *Ibid.*, p. 84.

58. Manes, *Ibid.*, p. 24.

59. *Ibid.*, p. 20.

60. *Ibid.*, pp. 40, 42, 133.

61. *Ibid.*, p. 71.

62. *Ibid.*, p. 62. Circumstances of 2020 have certainly borne this out. Science came to our rescue with COVID-19 vaccines available by December 2020. Meanwhile, in the midst of the 2020 pandemic, American officials insidiously discussed the do-nothing experimental possibility, in lieu of vaccinations, of establishing "herd immunity" via

causing younger-aged people (e.g., children) who evidently had less severe reactions to the virus, to become exposed and then passing it along to the more general population in any number of re-opened venues—thus creating untold difficulties upon swamped, hospitals overflowing with severe cases ... and having the potential to cause millions of deaths in the U.S.

63. Garrett, *Coming Plague*, p. 551.

64. *Ibid.*, p. 528. "Neither rat nor man has achieved social, commercial, or economic stability. This has been, either perfectly or to some extent, achieved by ants and by bees, by some birds, and by some of the fishes in the sea. Man and the rats are merely, so far, the most successful animals of prey. They are utterly destructive of other forms of life. Neither of them is of the slightest use to any other species of living things." Hans Zinsser quoted in Garrett, p. 528.

65. Manes discusses differences between Deep Ecology and New Age culture in *Green Rage*, p. 159.

66. Garrett, *Coming Plague*, pp. 577–578. Biologist Julian Davies cited therein.

67. *Ibid.*, p. 581. Dr. Bernard Fields cited therein.

68. *Ibid.*, pp. 618, 620.

69. *Ibid.*, p. 603. Karl Johnson cited therein.

70. *Ibid.*, p. 570. Also see Sunita Narain, "Why Climate Matters," *Time*, Vol. 197, nos. 23–24 (June 21/June 28, 2021), p. 95.

71. George A. Romero and Daniel Kraus, *The Living Dead* (New York: Tor, 2020), pp. 514–515.

72. Besides *Frankenstein*, Mary Shelley authored another science fiction novel in 1826, *The Last Man*—sole survivor of a pandemic. (See Debus, Chapter 35 in *Prehistoric Monster Mash*.) George R. Stewart (1895–1980) authored *Earth Abides* in 1949, concerning the Darwinian aftermath of a global plague, which award-winning science fiction author Connie Willis stated in 2005, "is usually classified as a 'post-apocalypse' novel, part of a group of science fiction novels and short stories written in the late forties and early fifties that chronicled the collapse of civilization and the struggles of a few scattered survivors and that

were at least partly prompted, if not by 'nuclear dread,' ... then by an uncomfortable post-Hiroshima awareness that humankind's residence on Earth might be only temporary." (Willis's Introduction to Stewart's *Earth Abides*, New York: Ballantine Books, 2006 Del Rey Trade ed., p. xiii.) Stewart's novel begins with a quote by W. M. Stanley from *Chemical and Engineering News* (Dec. 22, 1947), "If a killing type of virus should suddenly arise by mutation ... it could, because of the rapid transportation in which we indulge nowadays, be carried to the far corners of the earth and cause the death of millions of people." And the novel ends with biblical verse from *Ecclesiastes*, "Men go and come, but Earth abides" (p. 345). Stephen King has written many scary and supernatural disaster tales, of which *The Stand* (1978) may be most widely known, while Kirkman indoctrinated *The Walking Dead*'s immensely popular zombie apocalypse themed universe regarding an Earth relatively unpopulated—by the *living*, anyway—during the 2000s and 2010s. Frederik A. Stokes's 1889 novel *The Last American*, involving a post-apocalyptical world wreaked by climate change was among the earliest in this genre. (For more on this last title, unread by this author, see Merkl, *op. cit.*, p. 234.)

73. Invention of the hydrogen bomb set off a chain reaction of giant dino-monster (and huge radioactively charged bug) movies during the 1950s. Allen A. Debus, *Dinosaurs Ever Evolving*, *op. cit.*, pp. 116–117; Debus, *Prehistoric Monster Mash*, pp. 256–265.

74. Lovelock, *Vanishing Face of Gaia*. Note that Gaia theory was conceived prior to Rachel Carson's apt ecological warnings. In her book, Garrett doesn't mention Gaia or Lovelock, while, in his, Gore emphasizes the spiritual nature of Gaia theory (while suggesting it may lie more so in the realm of fringe science). However, Peter Ward views Gaia as a "foundation of modern environmentalism." *The Medea Hypothesis*, p. 39.

75. *Ibid.*, pp. 9, 13. Lovelock comments here on the mindset that by simply adopting green technologies and strategies we can "save the planet": "The real Earth does not need saving. It can, will,

and always has saved itself, and is now starting to do so by changing to a state much less favorable for us and other animals. What people mean by the plea is "save the planet as we know it," and that is not impossible." pp. 18–19. As he warns, non-linear, accelerating factors already in flux may ultimately exacerbate climate changes on the horizon, creating positive feedback systems that we cannot possibly control or reverse.

76. *Ibid.*, pp. 88–89.

77. *Ibid.*, p. 28, 32,

78. *Ibid.*, pp. 91, 97, 174.

79. Gaia's systems are interwoven with all life's intricacies. And when it comes to modeling the Earth system, a very difficult-to-solve set of *non-linear* differential equations must be used. Ray Bradbury's "butterfly effect" was captured in his acclaimed 1952 short story, "A Sound of Thunder."

80. *Ibid.*, pp. 105–118; Prior to Fukushima, there were only two nuclear reactor accidents—Chernobyl in the Ukraine and New York's Three Mile Island. These incidents were certainly frightening and in Chernobyl's case deadly, but do not compare in scope with the Fukushima daiichi March 11, 2011, power plant disaster (footage of which may have proven influential behind opening scenes in Legendary's 2014 *Godzilla*, as well as Toho's *Shin Godzilla*), in wake of the Tohoku 9.0 Earthquake and resulting 30-foot-high tsunami. Radiation was spread locally and into the ocean and there were 20,000 deaths associated with the awful event. High radiation levels persisting in the area prevent citizens from returning to their property. In utter contrast, historically, there are untold tens of thousands of individuals whose deaths may be attributed in some fashion to the production and consumption of oil, coal and gas for consumption as fuel.

81. Mane, *op. cit.*, p. 220; Toho's Mothra movies of the 1990s in particular have railed against the dangers of deforestation. Landslides and erosion of topsoil are outcomes of deforestation—a theme captured in scenes where Mothra's large egg is seen eroding out of the ground into the sea during storms. Al Gore also warned about the persistent loss of good agricultural topsoil across America's Midwest region in his *Earth In the Balance, op. cit.*

82. Besides *Star Trek IV*, a number of in your face eco-themed/climate change catastrophe films and television shows have 'hit home' over the past half-century. Here's an interesting sampling of, for the most part, recent and relevant titles: *The Birds* (Universal, directed by Alfred Hitchcock, 1963); *The Day After Tomorrow* (Twentieth Century Fox, directed by Roland Emmerich, 2004); *Snowpiercer* (TNT television series, 2020–2021); *The Midnight Sky* (Anonymous Content/Netflix, directed by George Clooney, 2020); *2067* (Arcadia, directed by Seth Larney, 2020); *Love and Monsters* (Paramount, directed by Michael Matthews, 2020).

83. Garrett, *Coming Plague*, p. 555.

84. The term "biological chain reaction" was coined by Richard Preston in his 2002 book *The Demon in the Freezer* (New York: Random House), p. 48. In an editorial, J. D. Lees discusses the chain of events concerning setbacks in getting *Godzilla vs. Kong* to anxious audiences in *G-Fan* vol. 1, no. 130, Winter 2020, p. 5.

Appendix

1. Robert Hood and Robert Pen, *Daikaiju! Giant Monster Tales* (Australia, University of Wollongong: Agog! Press, 2005), pp. vi–vii.

2. Jeremy Robinson, *Project Maigo* (Breakneck Media, 2013), p. 54.

3. Jeremy Robinson, *Project Hyperion* (Breakneck Media, 2015), p. 267.

4. J. D. Lees, "What is a Kaiju?" *G-Fan* vol. 1, no. 78, Fall 2006, pp. 68–72.

5. J. D. Lees, pers. comm., May 12, 2018.

6. Jose Luis Sanz, *Starring T. Rex! Dinosaur Mythology and Popular Culture* (Bloomington and Indianapolis: Indiana University Press, 2002), pp. 116, 120–121.

7. *Ibid.*, pp. 108–109, 114–115.

8. *Ibid.*, p. 115. Furthermore, "...the degree of Godzilla's anthropomorphism increases in proportion to the extent of its alliance with humanity." p. 120.

9. Lees, *op. cit.*, p. 72.

10. Both Legendary's and Joe DeVito's literary and artistically portrayed Kongs are naturalistic. Both have an alluded prehistory ancestry. Comparing Legendary's and DeVito's, the main significant distinction, respectively is the relative sizes of each of their last kongs. So, on the surface, isn't Legendary's Kong more kaiju-like than DeVito's simply because of relative size? Legendary's version defies the force of gravity and biophysical scaling laws. Isn't greatly exaggerated size *the most* defining parameter of any kaiju contender?

11. Greg Keyes, *Godzilla: King of the Monsters* (London: Titan Books, 2019), p. 144.

12. *Ibid.*, p. 195.

13. *Ibid.*, p. 298.

14. For the full scoop, however, see Chapter Twenty-two in Debus, *Dinosaur Memories II.*

15. John Taine, *The Greatest Adventure* (1929), reprinted in *The Antarktos Cycle: Horror and Wonder at the Ends of the Earth* (Canada: Chaosium Book 6031, Aug. 1999), pp. 165–304. Allen A. Debus, *Dinosaurs in Fantastic Fiction*, pp. 50–51.

16. Arvid Nelson and Zid, *Skull Island: The Birth of Kong* (Burbank, CA: Legendary Comics, Dec. 2017).

17. This cartoon is available online at YouTube. Jeff Rovin, *The Encyclopedia of Monsters* (New York: Facts on File, Inc., 1989), pp. 12–13.

18. Allen A. Debus, *Prehistoric Monsters: The Real and Imagined Creatures of the Past That We Love to Fear* (Jefferson, NC: McFarland & Company, Inc., Publishers, 2010), p. 222.

19. For more on the science and paleoart of pelycosaurs (e.g., particularly *Dimetrodon*, *Edaphosaurus* and 'Naosaurus'), see Chapter Twelve (titled "Fin-tastic Mammals") in my 2002 book *Dinosaur Memories: Dino-trekking for Beasts of Thunder, Fantastic Saurians ... and other 'Prehistoria.'*)

20. Keyes, 2021, p. 267.

21. Arvid Nelson, *Godzilla Aftershock* (Burbank, CA: Legendary Comics, 2019).

22. Joe Fordham, "Pacific Rim," *Cinefex* no. 135, Oct. 2013, p. 80.

23. Irvine, *Pacific Rim*, p. 200.

24. Keyes, *Godzilla*, 2019, p. 61.

25. *Ibid.*, p. 91.

26. *Ibid.*, pp. 106, 192.

27. *Ibid.*, pp. 198, 204.

28. *Ibid.*, p. 206.

29. Sanz, *op. cit.*, p. 114.

30. Or possibly the constantly growing dino-monster central to Henry Kuttner's 1940 short story "Beauty and the Beast," discussed further in my 2006 book *Dinosaurs in Fantastic Fiction* (pp. 107–110).

31. For much more on this creature, see Chapters Two and Eight in my book *Prehistoric Monster Mash.*

32. Keyes, *op. cit.*, pp. 141–142.

33. *Ibid.*, p. 175.

34. *Ibid.*, p. 168.

35. *Konga*, Charlton, vol. 1, no. 23, Nov. 1965, illustrated by "Masulli + Rocke," and "Montes-Bache."

Bibliography

Archer, Steve. *Willis O'Brien: Special Effects Genius.* Jefferson, NC: McFarland, 1993.

Asimov, Isaac, and Pohl, Frederik. *Our Angry Earth.* New York: Tor, 1991.

Berner, Robert A. *The Phanerozoic Carbon Cycle: CO2 and O2.* Oxford: Oxford University Press, 2004.

Berry, Mark F. *The Dinosaur Filmography.* Jefferson, NC: McFarland, 2002.

Bogue, Mike. "In Search of King Kong vs. Godzilla: One Fan's Holy Grail." *G-Fan* 125 (Fall 2019): 44–48.

Bogue, Mike. "Kaiju Retrospective: Japan's *King Kong vs. Godzilla* Under the Monsterfying Glass," in *Scary Monsters* no. 123 (Fall 2021): 33–36.

Borenstein, Max, and Borenstein, Greg. *Godzilla: Awakening.* Burbank, CA: Warner Bros. Entertainment, Inc., with Legendary Comics, 2014.

Bowler, Peter J. *A History of the Future: Prophets of Progress from H. G. Wells to Isaac Asimov.* Cambridge: University of Cambridge Press, 2017.

Bramwell, Valerie, and Peck, Robert M. *All in the Bones: A Biography of Benjamin Waterhouse Hawkins.* Philadelphia: The Academy of Natural Sciences, 2008.

Brinkman, Paul D. *The Second Jurassic Dinosaur Rush: Museums & Paleontology in America at the Turn of the Twentieth Century.* Chicago: University of Chicago Press, 2010.

Brown, Barnum. "Tyrannosaurus, a Cretaceous Carnivorous Dinosaur" *Scientific American.* Vol. 113, no. 15 (Oct. 9, 1915): 322–323.

Cox, Greg. *Godzilla.* London: Titan Books, 2014.

Czerkas, Stephen. "O'Brien vs Dawley: The First Great Rivalry in Visual Effects." *Cinefex* 138 (July 2014): 13–27.

Czerkas, Stephen A. *Major Herbert M. Dawley: An Artist's Life—Dinosaurs, Movies, Show-Biz, & Pierce-Arrow Automobiles.* Blanding, UT: The Dinosaur Museum, 2016.

Czerkas, Stephen A. "A Reevaluation on the Plate Arrangement on Stegosaurus stenops." *Dinosaurs Past and Present* Vol. II, Seattle: Natural History Museum of Los Angeles County in association with University of Washington Press (1987): 82–99.

Dean, Dennis R. *Gideon Mantell and the Discovery of Dinosaurs.* Cambridge: Cambridge University Press, 1999.

Debus, Allen A. *Dinosaur Memories II: Pop-Cultural Reflections on Dino-Daikaiju & PaleoImagery.* Amazon Createspace, 2017.

Debus, Allen A. *Dinosaurs Ever Evolving: The Changing Face of Prehistoric Animals in Popular Culture.* Jefferson, NC: McFarland, 2016.

Debus, Allen A. *Dinosaurs in Fantastic Fiction: A Thematic Survey.* Jefferson, NC: McFarland, 2006.

Debus, Allen A. *Prehistoric Monster Mash: Science Fictional Dinosaurs. Fossil Phenoms, Paleo-pioneers, Godzilla & Other Kaiju-saurs.* Kindle Publishing, 2019.

Debus, Allen A. *Prehistoric Monsters: The Real and Imagined Creatures of the Past That We Love to Fear.* Jefferson, NC: McFarland, 2009.

Debus, Allen A. "Reflections of Doomsday: When Did Dinosaurs Become Apocalyptical?" *Monster!* 27 (March 2016): 95–99.

Desmond, Adrian J. *Huxley: From Devil's Disciple to Evolution's High Priest*. Reading, MA: Perseus Books, 1994, 1997.

Du Chaillu, Paul. *Explorations and Adventures in Equatorial Africa*. London: John Murray, 1861, Ostara Publications reprint, 2014.

Glut, Donald F. *Classic Movie Monsters*. Scarecrow Press, 1978, Chapter VIII "His Majesty, King Kong": 282–373.

Glut, Donald F. *The Dinosaur Scrapbook: The Dinosaur in Amusement Parks, Comic Books, Fiction, History, Magazines, Movies, Museums, Television*. Secaucus, NJ: The Citadel Press, 1980.

Glut, Donald F. *Dinosaurs: The Encyclopedia* (Vol. I) (Jefferson, NC: McFarland, 1997), separate entries for *Stegosaurus* (pp. 840–853), *Tyrannosaurus* (pp. 945–960) *and Iguanodon* (pp. 490–500).

Glut, Donald F. *Dinosaurs vs. Apes* (2007, DVD) containing films "Hollywood Goes Ape!" and "Dinosaur Movies."

Godziszewski, Ed. "The Making of Godzilla." *Japanese Giants: The Magazine of Japanese Science Fiction and Fantasy* 10 (Sept. 2004): 4–47.

Golden, Christopher. *King Kong*. New York: Pocket Star Books, 2005.

Gould, Stephen Jay. "Knight Takes Bishop," Chapter 26, *Bully For Brontosaurus: Reflections in Natural History*. W. W. Norton & Company, 1991.

Gribbin, John, and Gribbin, Mary. *He Knew He Was Right: The Irrepressible Life of James Lovelock*. London: Penguin Books, 2009.

Horner, John R., and Gorman, James. *Digging Dinosaurs*. New York, Workman Publishing, 1988.

Horner, John R., and Lessem, Don. *The Complete T. Rex*. New York: Simon & Schuster, 1993.

Ilja, Nieuwland. *American Dinosaur Abroad: A Cultural History*. Pittsburgh, PA: Pittsburgh University Press, 2019.

Irvine, Alex. *Pacific Rim*. London: Titan Books, 2013.

Joseph, Lawrence E. *Gaia: The Growth of an Idea*. New York: St. Martin's Press, 1990.

Keyes, Greg. *Godzilla: King of the Monsters*. London: Titan Books, 2019.

Kolbert, Elizabeth. *The Sixth Extinction: An Unnatural History*. New York: Henry Holt and Company, 2014.

Kump, Lee R., Kasting, James F., Crane, Robert G. *The Earth System, 3rd Edition*. Uttar Pradesh, India: Pearson Education, Inc., 2016.

Lebbon, Tim. *Kong: Skull Island*. London: Titan Books, 2017.

Lees, J. D., and Cerasini, Marc. *The Official Godzilla Compendium*. New York: Random House, 1998.

LeMay, John. *Writing Japanese Monsters*. Roswell, NM: Bicep Books, 2020.

Lovelace, Delos W. *King Kong*. New York: Grossett & Dunlap, 1933, New York: Random House, The Modern Library Classics, 2005.

Lovelock, James. *The Ages of Gaia: A Biography of Our Living Earth*. W.W. Norton & Company, 1988.

Lovelock, J.E. *Gaia: A New Look at Life on Earth*. Oxford University Press, 1979.

McGhee, Jr., George R. *Carboniferous Giants and Mass Extinction: The Late Paleozoic Ice Age World*. New York: Columbia University Press, 2018.

Meier, Gerhard Meier. *African Dinosaurs Unearthed: The Tendaguru Expeditions*. Bloomington: Indiana University Press, 2003.

Merkl, Ulrich. *Dinomania: The Lost Art of Winsor McCay, the Secret Origins of King Kong, and the Urge to Destroy New York*. Seattle: Fantagraphics Books, 2015.

Mitchell, Ashley. "Godzilla: The Deity of Destruction," *G-Fan* vol. 1, no. 131 (Spring 2021): 41.

Mitchell, W. J. T. *The Last Dinosaur Book: The Life and Times of a Cultural Icon*. Chicago: University of Chicago Press, 1998.

Nelson, Arvid. *Godzilla: Aftershock*. Burbank, CA: Legendary Comics, 2019.

Nelson, Arvid. *Skull Island: The Birth of Kong.* Burbank, CA: Legendary Comics, 2017.

Norman, David. *Dinosaur!* New York: Prentice Hall, 1991: 160–169.

Pettigrew, Neil. *The Stop-Motion Filmography: A Critical Guide to 297 Features Using Puppet Animation.* Jefferson, NC: McFarland, 1999.

Preston, Richard. *The Demon in the Freezer: A True Story.* New York: Random House, 2002.

Rainger, Ronald Rainger. *An Agenda for Antiquity: Henry Fairfield Osborn and Vertebrate Paleontology at the American Museum of Natural History, 1890–1935.* Tuscaloosa: University of Alabama Press, 1991.

Reel, Monte Reel. *Between Man and Beast: An Unlikely Explorer, the Evolution Debates, and the African Adventure That Took the Victorian World by Storm.* New York: Doubleday, 2013.

Rovin, Jeff. *From the Land Beyond Beyond: The Films of Willis O'Brien and Ray Harryhausen.* New York: Berkley Publishing Corp., 1977.

Russo, James. "Korean Kaiju: Symbolism in *Yongary* and *Pulgasari.*" *G-Fan* 130 (Winter 2020): 64–69.

Ryfle, Steve. *Japan's Favorite Mon-Star: The Unauthorized Biography of 'The Big G.'* Toronto: ECW Press, 1998.

Sagan, Carl. *The Dragons of Eden: Speculations on the Evolution of Human Intelligence.* New York: Ballantine Books, 1977.

Sanz, Jose Luis. *Starring T. Rex! Dinosaur Mythology and Popular Culture.* Bloomington: Indiana University Press, 2002.

Tsutsui, William. *Godzilla on My Mind: Fifty Years of the King of Monsters.* New York: Palgrave/Macmillan, 2004.

Turner, George E., and Goldner, Orville. *Spawn of Skull Island.* Baltimore, MD: Luminary Press, 2002.

Verne, Jules. *A Journey to the Center of the Earth.* (1872 English translation—"West Corner of St. Paul's Churchyard": Griffith and Farran; Secaucus, NJ: Signet Classics ed., 1986).

Ward, Peter D. *The Medea Hypothesis: Is Life on Earth Ultimately Self-Destructive?* Princeton, NJ: Princeton University Press, 2009.

Ward, Peter D. *Out of Thin Air.* Washington, DC: Joseph Henry Press, 2006.

Ward, Peter D. *Under a Green Sky.* New York: Smithsonian Books-HarperCollins Publishers, Inc., 2007.

Woods, Paul A. ed., *King Kong Cometh.* London: Plexus Publishing, Ltd., 2005.

Worland, Justin. "Standing at a Climate Crossroads, The Last Exit Before Catastrophe," *Time* vol. 196, nos. 3–4 (July 20/July 27, 2020): 34–44.

Index

Numbers in **bold italics** indicate pages with illustrations

245